introduction 简介

interviews 访谈

architects 建筑师事务所

introduction 简介

Bart Goldhoorn

FREEING THE FLOOR PLAN 自由式设计

Freeing the Floor Plan
The Rise of Interior Design in Russia
Bart Goldhoorn

自由式设计
——俄罗斯室内设计的崛起
巴特•高德霍恩

中文摘要:

我亲历俄罗斯室内设计的发展始于1993年，当时我在荷兰文化部授权批准下来到了俄罗斯与一群年轻的建筑师一起工作。那时的俄罗斯由于受到体制的限制，建筑和室内设计的发展陷入窘境。建筑以实用为主，几乎谈不上设计。15年后，情况发生了改变，俄罗斯室内设计的鼎盛时期到来了，个性化的室内设计出现，紧追国际设计发展潮流。尤其特别指出的是俄罗斯自由式公寓设计，它不仅是自从1980年代纸面建筑之后俄罗斯向国际建筑领域的第一份重要的贡献，也是俄罗斯室内设计走向繁荣的原因之一。俄罗斯绝对数量的室内设计作品已经导致了大部分建筑团体活跃在这个领域。这意味着不仅可以在自由设计公寓中看到这些成果，也可以在那些旧式建筑公寓、餐馆、俱乐部和精品店中欣赏到这种设计的辉煌。相比而言，俄罗斯“真正的”建筑看起来比较黯淡，这可能是俄罗斯文化特性的结果。总之，室内设计仅仅是个场景的布置——它不可能像建筑物一样长久存在，但是它更符合俄罗斯艺术、文学、戏剧和电影领域创造的亦真亦幻的传统成就。

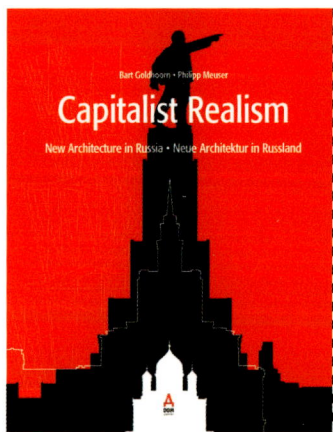

Goldhoorn/Meuser:
*Capitalist Realism.
New Architecture in Russia,*
DOM publishers, 2006

My personal experience with interior design in Russia started in 1993, when I came to Russia on a grant from the Dutch Ministry of Culture and worked together with some young architects that had just started their own architectural firm. At that time, nobody had money and virtually nothing was being built in Moscow. The only realistic clients were small businessmen who wanted their shops or showrooms fitted out. The work of the architect in this situation didn't have too much to do with design, let alone beauty – his main task was to be sure that at least something was built. In the absence of small construction companies and a transparent market for building products, architects found themselves organising a »brigada« and roaming the markets trying to find building materials. Often the choice for a certain material was dictated by availability rather then considerations of design. Thus it could happen that a shop interior was completely painted in metallic blue since this was the only colour the architect had been able to get hold of. Another example: when a small window was broken in our office, the handyman that did odd jobs for us at the time came back with a German do-it-yourself package for framing pictures he had happened to find in a shop around the corner. This was the only possibility to actually get a small piece of glass in Moscow – the only other way was to steal it or to order a truckload.

This illustrates the situation in architecture in Moscow at the beginning of the 1990s, and even more so the situation in interior design. Constructing buildings was an activity that was more or less compatible with the mechanisms of communist planning that were still operational by reasons of inertia. Interior design asked above all for a free market – a place where one can find a variety of different products to suit the taste of the individual consumer. And a market was exactly what was missing in Russia at that time.

15 years later the situation is practically the opposite – if architecture still suffers from the bad quality of design and construction inherited form the Soviet period, interior design is blooming and has caught up with international developments. Russian interiors are published in foreign magazines, Russian designers are flocking the *Milan Salone di Mobile*, Moscow's streets are dotted with foreign interior showrooms and yes, the first monography on Russian interior design (that is, this one) is published.

In order to enable architects to find materials, of until 2005, each issue of *Project Russia* contained a catalogue of building products.

The Birth of the Interior Architect

One of the reasons is that talented young architects in the beginning of the 1990s got involved in interior design – they could not find employment in architecture. Architecture was (and partly still is) controlled by the older generation of architects, who are well connected in the former communist »nomenklatura«. As in any authoritarian system, age is the main prerequisite for getting something done. Older people know how to deal with the regulations and norms and unwritten rules that are governing the construction process – in the end it are the people of their own generation that are pulling the strings.

Interior architecture offered an alternative to the perspective of slowly climbing up in the hierarchy of the project institute, trying to get your project approved by an endless row of bureaucrats. In contrast, interior design offered independence and instant success, and in this sense interior design became a school of architecture for a whole generation of Russian architects. Some of them still work in this field, some of them have moved on and are now actually practicing architecture. One of the biggest differences of the interior with architecture

practice was the enormous freedom of the architect. In the Russian construction industry, architecs play a minor role since Khrushchev's reforms of the 1960s, and to a certain extent this is still the case. This freedom and independence also brought responsibilities. Architects would not only design the interiors, but also build them. In fact they became contractors, and they were forced to develop practical knowledge of materials and construction methods, when the market for interior fittings was still very underdeveloped. Often architects would design custommade details instead of just assembling standard elements. This was made possible by the availability of highly qualified craftsmen that would work for relatively low wages. These craftsmen could be found in very disparate fields: on the one hand you had artists with a classical schooling in the Soviet art schools, on the other hand the Soviet high tech defence industry that was desperate for work. There are many examples of furniture or lamps designed by architects and produced by companies that were involved in the Soviet space industry.

An exceptional case is Bioinjector, set up by Igor Safronov, an engineer that had been working on the Soviet space shuttle. After an attempt to produce a high-tech medical instrument for injecting insulin

9

NEW INTERIOR DESIGN IN RUSSIA

Igor Saronov's bank safes

failed because he couldn't find a market for them, in the beginning of the 1990s, he decided to produce a product he was sure would sell – bank-safes. He invited some young architects to make the designs, and as a result he was – and still is – producing the only design safes in the world.

His contact with the architecture world led to the development of another line of work – the production of custom-made furniture and interior fittings. The works produced by his firm and other small metal workshops (often started by former employees) can be found in many works by young Russian architects of that time. The most extreme example are the works by Alexey Kozyr, such as his »airplane apartment« – a 30 square metres one room apartment that is fitted out with a door made of an old Soviet bomber, a bed with a removable steel bridge used to reach the balcony and a bathtub with an aquarium around it.

However, one didn't have to go into high-tech to get hands-on experience with realizing ones projects. Any architect working in interior design had to deal with material supplies, project management, working drawings and explaining builders what he wants them to do. This is even the case with the most popular material in interior design: gypsum board.

Notwithstanding the practical skills needed to realize it, this is where the interior comes closest to its primary purpose as a décor for living. In this way Russian architects were very well prepared to this field of work. In the 1980s, the fact that architects were cut off from the construction process led to the emergence of the Paper Architecture movement: young Russian architects won many international competitions with their beautifully drawn architectural fantasies. These works are not primarily representations of buildings, with floor plans, elevations and sections – they are works in itself, objects of art that require a direct involvement with material, colour and texture. The step from paper architecture to gypsum board architecture was not that big. Many architects from that movement got involved in interior design. A special mention must be made here of Mikhail Filippov, who managed to realize his incredible Piranesi-like architectural fantasies on the scale of apartment interiors.

In 1998, a group of well-known interior designers established a society called *Moscow Architecture Society* (MAO) – a name lent from the organisation thad had existed until 1939 and that included well-known constructivists like Ginzburg, Melnikov and Golosov. The aim of the new MAO was to establish an

Alexey Kozyr's
»Airplane Apartment«

organisation that would serve the interests of professional interior designers and discuss problems faced by the profession at the time when professional organisations in the field of architecture such as the Moscow and Russian Unions of architecture completely ignored this field. The older generation of architects saw interior architecture as something superficial that shouldn't be taken too serious. At the same time interior architects were making the most money – MAO established its own rate at 400$ per square metre. Design costs – much higher then the prices that were to be received by their colleagues doing 'real' architecture, even not taking into account the fact that interior designers get percentages of the furniture they buy for their clients.

The Demand for Interior Desgin

So if interior designers were so successful, where did the demand come from? In order to understand this, one should be aware of the situation on the Russian housing market. Under communism, no private houses had been built in Russian cities. Even the members of the Politbureau lived in apartments. Moreover, the majority of the apartments were built after Khrushchev's reforms of the 1960s, meaning

that they were constructed on the basis of standard floor plans that were virtually the same for the whole country. If the floor plans of the first generation of these buildings were small but still more or less OK architecturally – they were designed by architects who believed they were finally able to solve the housing problem in Russia – the later ones deteriorated under influence of the power of the building industry. This is the Soviet reality where everybody, including the wealthy oligarchs came from, and this is the reality from which they hope to escape. The contemporary Russian wants something unique and personal and is ready to pay a lot of money for this.

It is telling that the same does not seem to apply to the buildings the apartments and offices of these new Russians are located in. There is a big difference in attitude towards private and collective space. At the root of this attitude lies again, Soviet history: Soviet propaganda monopolized anything that had to do with the collective. Since architecture belongs to collective space, no high value is put upon it. Anything outside of the apartment building is generally seen as something alien that doesn't really concern the inhabitants. Building budgets are generally very low, materials used are cheap. In contrast, budgets for fitting out the private interiors are almost limitless.

apartment by
Mikhail Filippov next to one
of his architectural fantasies

Why an interior architect?

The question that still remains unanswered is why you would actually need an interior architect. Although the demand for design furniture is also high in Europe, this doesn't mean that it is bought by interior architects. Consumers buy their furniture and interior decorations themselves, whereas for simple adaptations there exists a whole DIY industry enabling consumers to adapt their interior to their wishes by their own hands.

There is a number of reasons why the situation in Russia is different. First of all, it is the relative inexperience of Russian consumers with making choices. Well known is the story of the Soviet citizen who gets in a complete shock when he is confronted with the possibilities of choice in a Western supermarket. Although this is of course not the case with contemporary Russian citizens, some of this insecurity has remained. And if you feel uncomfortable in deciding what you want, you ask a professional designer to do it for you. Another factor related to the lack of choice in Soviet time is the absence of inherited or earlier bought furniture. In general, Soviet furniture was of bad quality and design, and when people can afford to move into a new apart-

ment they will bring nothing with them – they start from scratch. The interior architect helps them to reinvent their life style. In many apartments in Moscow that are bought as second or third homes this factor is even more important – people will not live there very often and their purpose is more close to a hotel then a personalized space.

Another big difference with Western Europe is that notwithstanding the high fees of the interior designer, the difference between high and low income in Russia are enormous. Once you can afford to buy a new apartment, you are not going to do any renovations yourself but you will hire some workers (mostly immigrant workers from the former Soviet republics) and a designer to tell them what to do. In a European context, with the majority of the population belonging to the middle class, hiring workers is much more expensive and people tend to do much more themselves, including decisions about decoration and layout.

Probably the main reason why people need an interior designer is the fact that there is the earlier mentioned lack of variety in the housing they can buy in the city.

Firstly – the city itself is very homogenous. Apart from a small historical centre most of it con-

group portrait of MAO, 1998

sists of the standard apartment buildings that have been constructed since the 1960s. Compared to European cities, where you can choose between neighbourhoods with buildings from different eras, various social constellations and housing typologies, in Russia different parts of the city barely differ from each other. A complicating factor is that Russians don't trust the technical state of old buildings. In the Soviet Union old buildings were neglected and new buildings were presented as the results of technological progress, meaning that they were by definition better.

Secondly – all people live in apartments. If they have a house it will be outside the city and in general be a second home. In the big Russian cities you will hardly find any private houses, and if you do, they are generally wooden houses that are falling apart. Russian cities were almost completely constructed under Soviet rule, and Soviet urban design was based on the construction of apartment buildings surrounded by public space. Existing private houses were not taken care of – many of them even now have no proper sanitary and heating. Consequently, private houses in the city have a very bad name, as does living on the first floor. Private and collective space come too close.

If the supply of housing is limited to apartments, the architecture of the building has very little influence on the interior layout. In general, living in a private house will mean living on more then one floor. The limited size of each floor, the location of the stairs and windows dictates the use of the space. The architect that designed the house has for a large part decided how the space looks and how it can be used. This also means that the house already has a certain character that makes it different from others. An apartment layout is much less predetermined and consequently is much more anonymous.

The uniformity in both location and typology leaves consumers with not much choice in buying a place to live that is unique and with which one can identify. Then the only thing you can do in order to make your anonymous apartment into something unique and special is to invite an interior architect.

Free plan

Actually, the demand for individuality has led to a situation where in Moscow all expensive apartments are built with interior walls. »Svobodnaya planirovka« – free plan – is the term that is used to advertise these apartments. To architects this

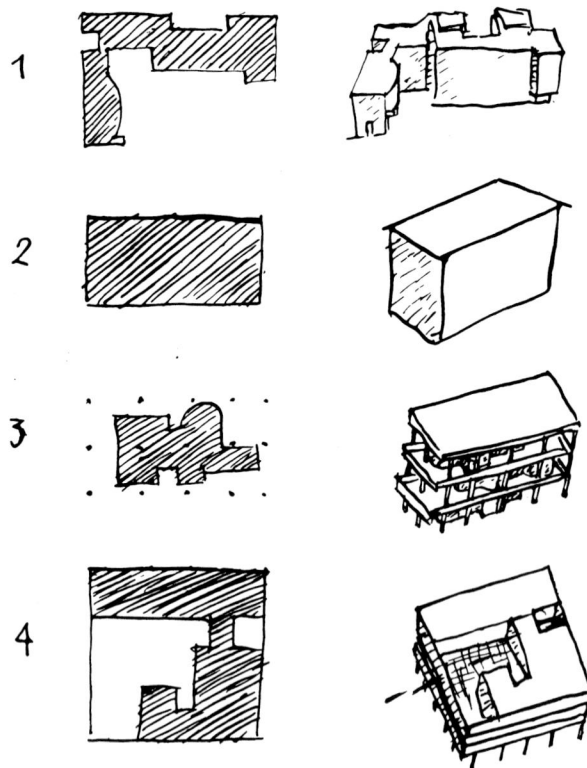

Le Corbusier, 1930, sketch

1. Maison
LaRoche-Jeanneret (1924)
2. Villa Meyer/Stein
de Morzine (1926)
3. Villa Baiseau II (1929)
4. Villa Savoie (1929)

TPO Reserv
apartment block in
Moscow, 2000

plan of the building
plans of the apartments

sounds familiar – wasn't this the term Le Corbusier used in his »5 points«: Le plan libre – with its interior walls independent of the column structure? At that time, this architectural concept was meant for internal use only: to free the architect from the harness of the load-bearing walls, enabling him to make free flowing spaces. It was not until the late 1960's that the idea occurred that this concept could actually be used to give the inhabitant of the dwelling more to say about his environment. Architects would develop »construction kits«, giving the inhabitants ample choices of room-sizes, placement of doors and even windows. Notwithstanding the unmistakably political agenda there was no question of giving up the position of the architects as the one to emancipate the people. Actually, an organization of architects with the name »Open Building« still exists, promoting the principle of the division of a »bearer« and »infill«.

So what is propagated by the idealist architects in Europe and America is realized by project developers in Moscow that don't have the least intention to improve the world but just follow the demands of the market. When they started to develop apartment complexes for the new rich in the beginning of the 1990's, they found out that the moment they would hand over the key to the new owner con-

struction workers would move in to remove not only kitchens and bathroom fittings, but also all interior walls. Seventy years of Soviet housing design had led the clients to the belief that any apartment layout that was not custom made would be not fit their individual taste or life style. Once developers understood this, they built all their apartments without interior layout.

Actually, the Russian model of shell and core apartment construction can be considered a real innovation in architecture: it finally has solved the problem of collective design and individual use that has been a central theme ever since mass housing has started to occur. It realizes on a big scale what has been tried in the West only in experimental projects. Of course this solution has a price: In Moscow, the price of an interior fit out including furniture can be close the price of the apartment proper. Also, the freedom to realize one's own interior might very well be limited to the first user. Considering the investment made by the first owner, the price of a finished apartment will be much higher than an empty one, meaning that it will be less opportune to destruct and replace it by a lay-out of one's own.

In any case the free plan principle leads to the solution of the problem of diversification: If all apart-

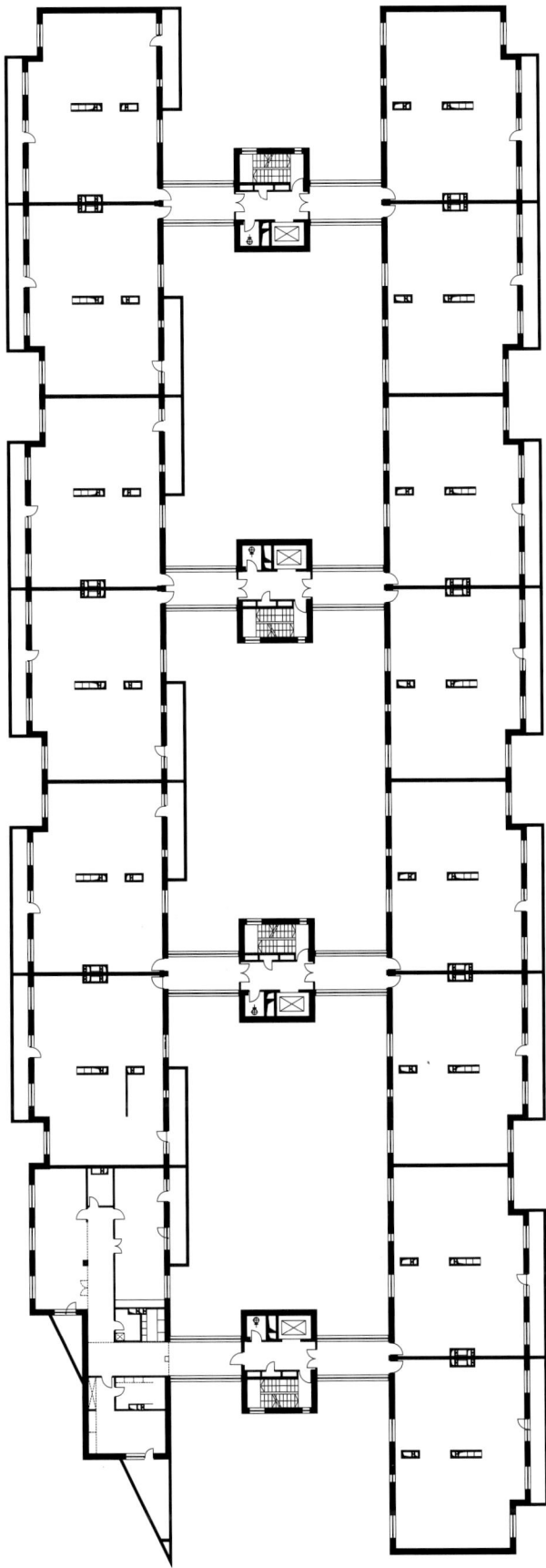

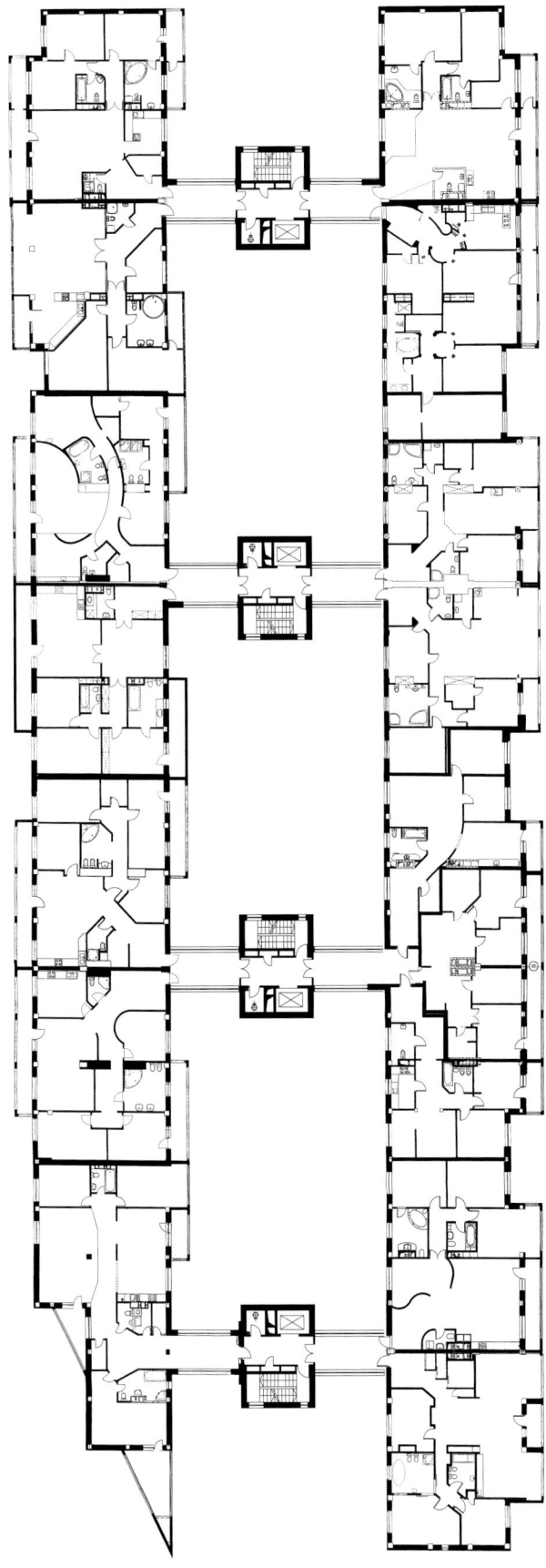

15

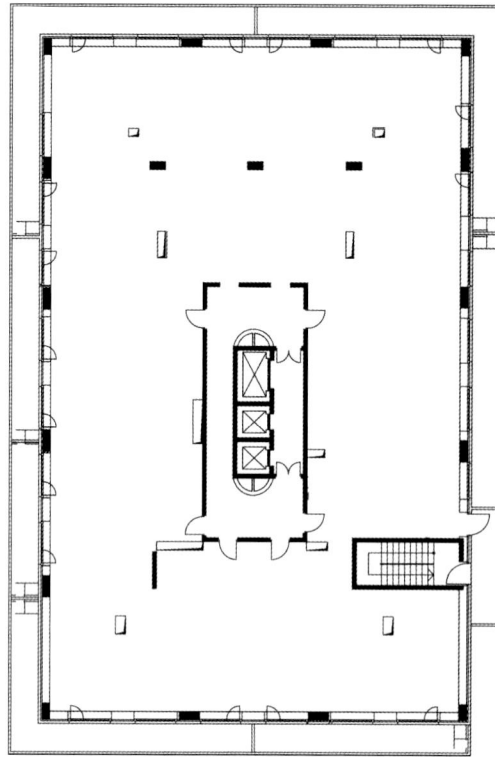

Ostozhenka Architects,
Sokol apartment building,
floor plan
Moscow, 2001

Ostozhenka Architects,
Sokol apartment building,
Moscow, 2001

ment interiors are designed by different architects, the consumer will have an enormous choice of options, one of which will certainly fit to his taste. In fact it can lead to a market for apartments that will have more similarities to the art market then to the market for real estate. The name of the designer will decide the price of an apartment interior as much as the name of the artist is decisive for the price of a work of art.

Not only is Russian free plan apartment the first serious contribution of Russian architecture to the international architectural debate since the Paper Architecture of the 1980s, it is also the reason for the flourishing of Russian interior architecture. The sheer amount of work in interior design in Russia has led to the fact that a large part of the architectural community is working in this field. This means that the results can not only be seen in the apartments with a free plan, but also in other interiors: those for apartments in older buildings, for restaurants, clubs and boutiques. That in comparison »real« architecture in Russia looks very bleak, is probably the result of the particularities of Russian culture: presentation is always better then performance. In

the end, the interior is a stage set – it might not have the same permanence as architecture, but it fits much better in Russia's tradition of creating mesmerizing illusions in art, literature, theatre and film.

interviews 访谈

Philipp Meuser

A positive mood dominates Russian architecture

Interview with Anton Nadtochy and Vera Butko

如今乐观的情绪在俄罗斯建筑领域占据着支配地位

建筑师夫妇安东•纳德托基（Anton Nadtochy）和维拉•卜特高（Vera Butko）访谈

The architect duo Anton Nadtochy and Vera Butko works under the name ATRIUM and extremely close to Western European-characterized design traditions addressing simplicity and conceptual coherence. In their designs, Nadtochy/Butko try to grasp light and material as a compositional basis, while referring not only to Soviet Constructivism of the 1920s but also to the philosophical deconstructivism of French postmodernism. The ATRIUM office strictly follows an all-encompassing approach, and its founders vehemently reject the term *design*. For them, every object qualifies as architecture.

When I look at your projects and at those of your colleagues in Russia, specific questions come to mind: Has a local, interior design language developed since the end of the Soviet Union? Is there a so-called Russian interior design?

AN: We're of the opinion that no specifically Russian interior design exists. Even when the situation concerns interior design, we're actually talking about architecture. There is no Russian interior design tradition for our work to actually refer to. Nevertheless, we surely have a few roots we can identify with. I mean the Russian avant-garde. That's an architectural tradition we would classify as Russian, and it clearly influences our work.

If you reject the notion of there being a Russian interior design, what do you refer to while completing your interior design projects? Are certain traditions, historical references, and philosophical-theoretical approaches present in your projects?

VB: We would call our work method experimental. It concerns conducting experiments with space. Beginning with an analysis of the space, it ideally leads to an overall concept, to a so-called theory of this space. But maybe that sounds abstract. I'll try phrasing it differently: We don't want to design any of the objects in the space; nor do we want to create any decorations. We want to design the actual spaces.

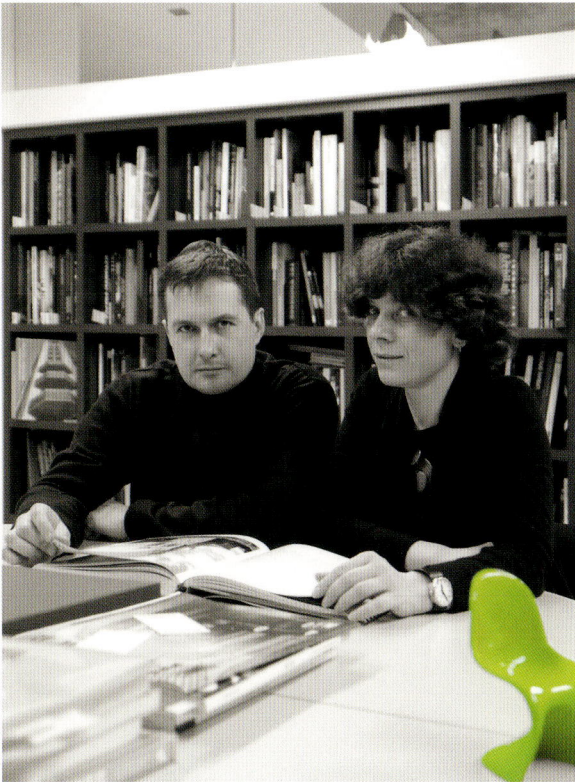

Anton Nadtochy
born 1970, grew up in a family of architects. In 1992 he completed his studies at the *Moscow Institute of Architecture* (MARCHI) and afterwards worked on a dissertation devoted to contemporary architecture. He is a member of the Architect's Association.

Vera Butko
born 1965, studied at MARCHI as well. Her 1987 dissertation was awarded the Grand Prize of the Degree Projects' Competition, which included the collected Soviet Union submissions. Currently a member of the Architect's Association, Butko has won numerous international competitions. Anton Nadtochy and Vera Butko founded the ATRIUM office in 1994 and have completed over 40 projects since then.

安东·纳德托基
生于1970年，生长在一个建筑师的家庭里。1992年他完成了在莫斯科建筑学院（MARCHI）的学习生活，之后他继续钻研当代建筑专题。他是建筑师协会的会员。

维拉·卜特高
生于1965年，也在莫斯科建筑学院就读。1987年她的学位论文获得了学位设计竞赛特奖，此次大赛的范围涵盖整个前苏联设计作品。当前，作为建筑师协会会员的卜特高已经获得了国际性竞赛的多项大奖。1994年安东·纳德托基和维拉·卜特高创建了ATRIUM工作室，自从那时起他们已经完成了40余项设计。

Indeed, that sounds abstract. If you conduct "experiments with the space," it would interest me to know how you perceive architectural space...

AN: For us, this is the space existing between volumes and objects. The difference between architecture and interior design lies in the abiding condition. Interior design has to determine, define and design the space between objects or items. In contrast, the task of architecture lies in creating a building, which represents an object in and of itself. What we work with is the resulting space formed by architecture's massive, immobile structures, the space that opens up between these structures. Actually, this is a void, an empty space.

If I follow you correctly, the architectural space is contained by surrounding walls. Do these containing and limiting aspects play a role in your work, that is, in establishing a boundary between an interior and exterior? Or do you equate this with the human body? Its physicality, its anatomy, protected by clothing. Does this dividing layer enter into your work?

AN: How this skin is handled changes from project to project. The starting point is always the existing situation. This usually concerns handling concrete. Over and over again, the goal is to find a new quality through the transformation of this element.

VB: If there's a beautiful view, for example, and it remains sealed away by a windowless wall, we try to break through the wall so that this quality can also be experienced inside. In our designs, this skin's material plays a role as well. Whether it concerns concrete or brick, we always try to integrate this quality, the wall's optic and haptic presence, into the overall concept and exploit it.

You're sitting before a huge wall of book shelves filled with architectural volumes. How has architectural theory influenced your work?

AN: I studied architectural theory and history very intensively. That occurred during a period when I couldn't work as an architect, because young architects were never granted contracts then. That was during the early 1990s. This was nevertheless an extremely valuable period for me. I was able to think things through to the end and develop my own architectural position. The approach that I used then is still important for me now: urban planning, architecture and interior design should always be conceived in unison.

21

Cappellini furniture boutique
2001

VB: Theory is still extremely important for us today, if less so as theorists and more so as practicing architects. We read books, we travel extensively, and we attend exhibitions. These activities offer the best opportunities for an architect's further development. This need for information evolved from a period when Russia was practically cut off from all international information sources. Even now we try to somehow compensate for that. Books in particular offer us a tremendous help; they teach us to learn from our mistakes. Today's developments occur so quickly that we need the different media in order to stay up to date, regarding either the latest trends or technical innovations.

We invited you to an exhibition under the heading "Lust auf Raum" (In the Mood for Space). What do you expect from an exhibition with a title like that?

AN: I think the heading is very appropriate. The expression 'In the Mood for...' perfectly describes the atmosphere in Russian architecture today – an atmosphere dominated by an optimistic mood nowadays. 'Space' is, of course, an established concepts in which meanings, interpretations and perceptions can be articulated. Space is permanently open to experiments. My expectations amount largely to

my curiosity: I'm curious about my colleagues' work.

VB: You can always make striking comparisons on occasions like these. Afterwards, people might even be able to say whether there really is a Russian interior design.

维拉·卜特高：我们愿意把我们的工作方式称为实验性的。这涉及到利用空间做试验。首先分解空间，它非常理想地指向一个总的概念，也就是所谓的空间理论。可能听起来有些抽象。我将试着用不同的语言表述这个意思：我们不想设计任何存在于空间中的物体，我们也不想创造任何装饰。我们想要设计实际存在的空间。

实际上，听起来很抽象。如果您用"空间做试验"，我非常感兴趣的是您如何感知建筑空间……

安东·纳德托基：对于我们来说，空间存在于体积与物体之间。建筑与室内设计之间的差异性存在于永恒不变的环境之中。室内设计必须确定、定义和设计物体间或者个体间的空间。相反，建筑的任务在于正在建造的一栋建筑中，表现一个物体在其本身内和本身具有的属性。我们工作的空间是由建筑结实稳定的结构构成的结果空间，这一空间在结构间展开。实际上，这是一个什么也没有的空间。

您正坐在一面巨大的满是建筑书籍的书架墙前面。那么建筑理论是如何影响您的工作的呢？

安东·纳德托基：我非常饥渴地学习建筑理论和历史。这种情况出现在一个时期内，那时我还不能成为一名建筑师，因为年轻的建筑师那时是从来都不会被聘用的。那是在20世纪90年代早期的时候。但是对于我这绝对是一个非常有价值的时期。我能够彻底的思考事情，并且找到了我自己在建筑领域的定位。那时使用的方法对于现在的我仍然很重要：城市的规划、建筑和室内设计应该需要和谐地构思。

维拉·卜特高：今天对于我们来说理论仍然是相当重要的，如果过少就是空谈家，过多就是从业的建筑师。我们读书，到处旅行，参加展览会。这些活动对于一名建筑师的进一步发展提供了最佳的机会。对信息的需求来源于一个时期，当时俄罗斯实际上被切断了与外界信息的所有联系。甚至现在我们都在设法以某种方式补偿那种缺失。特别是图书提供给我们巨大的帮助；它们教会我们从错误中学习。今天的发展是那么的迅速，我们需要不同的媒介物与时代同进步，也许是最新的流行趋势或者是技术的变革。

我们邀请您参加一个题为"情绪空间"的展览会。您期望从这个展览会中获得些什么呢？

安东·纳德托基：我认为这个题目非常的恰当。"情绪……"非常棒地描述了俄罗斯今日建筑的氛围——如今一个乐观的情绪统治的氛围。当然，"空间"是一个被建立的概念，在这个概念中含义、解释和理解可以被清晰地表达。空间对于试验是永远开放的。我的期望值很大程度上等于我的好奇心；我对同事的作品充满了好奇。

维拉·卜特高：在这样的场合中，你总是能做出一些惊人的对比。之后，人们甚至能够说是不是真正的存在一种俄罗斯式的室内设计。

访谈摘要：

建筑师夫妻档安东·纳德托基和维拉·卜特高在名为ATRIUM的工作室中工作，设计风格非常接近欧美特色设计传统，表达了简捷和概念统一的设计理念。在他们的设计中，纳德托基和卜特高努力抓住光和材料作为布局的基础，涉猎的领域不仅包括1920年出现的苏联构成主义风格，也包括法国现代主义哲学解构主义风格。ATRIUM工作室严格遵循着全盘吸收的方法，它的创建人强烈地拒绝"设计"这个词语。因为对于他们来说，每件物体都具有建筑的品质。

当我看到您的设计项目和您的俄罗斯同事的设计作品时，一些特别的问题出现在我的脑海里：自从苏联结束后，存在一个已经发展的本土化的室内设计语言吗？存在一个所谓的俄罗斯式室内设计吗？

安东·纳德托基：我们认为确切的说不存在俄罗斯式的室内设计。即使当情况涉及到室内设计的时候，我们实际上正在讨论的是建筑。没有俄罗斯式室内设计的传统供我们实际工作参考。然而，我们确实拥有几个能够认同的设计之源。我指的是俄罗斯的先锋派。那是一个我们愿意归属俄罗斯的建筑传统，并且，很明显它影响着我们的工作。

如果您否认存在俄罗斯式的室内设计概念这一观点，那么当您在完成设计项目时参考的东西又是什么呢？呈现在您的设计项目中的是某些传统、历史参考资料还是哲学理论方法呢？

Russian interiors have to be festive and opulent
Interview with Boris Bernaskoni

俄罗斯室内设计必须是欢乐的和丰富的

建筑师鲍里斯•伯纳斯科尼（Boris Bernaskoni）访谈

The work of Boris Bernaskoni demonstrates an unusually broad range. It includes smaller projects such as those for furniture designs and built-in constructions but also large-scale, public buildings and the planning of vast urban areas. Yet all of Bernaskoni's designs possess their own unique clarity as well as a thoroughly ironic quality, which the young architect knowingly uses in order to separate himself from the mainstream. By putting great stock in details and functionality, his architecture does justice to both good style and a newly conceived, high standard of living.

The meeting takes places in a noisy café on ulitsa Tverskaya. Because traffic noise is mixing with boisterous customers and a »Best of the Classics« CD blaring in the background, we can barely hear ourselves speak. Grinning, Boris Bernaskoni orders a coffee.

Why did you choose, of all places, this one for our interview and not your quiet studio?

I chose this café by coincidence. In any case, I think it's better here than in the office. My studio is just a kind of tool for me, an aid for my work. Besides, all architects' offices look somehow alike, don't they?

Apart from that – do you think there really is a specifically Russian interior design?

Yes, I do. For me, however, interior design is connected to the aspect of rationality, which makes Russian interior design amazingly irrational, because it relies more on excess and beauty than it does on the laws of reason. This has as much to do with the people who create this type of interior design as it does with laziness. In Germany everyone refers to Neufert, a volume accessible since the 1930s; it represents the most important set of tools for architects. This ergonomic, reference book provides a clear framework for design work. But nothing of this sort is ever used in Russia. Here everything develops more arbitrarily, more by chance. First you have an idea. Then you trust your instincts without thinking about the details. In the end, all that matters is the aesthetic idea, the style. Judging on the basis of your opinion, you're nothing like your Russian colleagues who, when asked if

Boris Bernaskoni
born 1977 in Moscow, studied at the *Moscow Architecture Institute* from 1994 to 2000. His extracurricular course studies earned him an additional degree in marketing. He completed his studies with a master's class in architectural composition. From 2001 to 2004, he was responsible for numerous exhibitions. He published various books addressing themes such as architecture, design, and urban development. Included here is Architektur der Grenze (The Architecture of the Border, with S. Morey). Bernaskoni is among the founders and editors of magazines such as Architectural Materials and A3.

鲍里斯·伯纳斯科尼
1977年出生在莫斯科，1994年至2000年在莫斯科建筑学院学习。他利用业余时间学习，另外取得了一个营销学位。在建筑方面，他完成了硕士课程的学业。从2001年到2004年，他负责了大量展览业务。他出版了各种各样的图书，涉及到各类主题，例如建筑、设计和都市发展等。其中包括Architektur der Grenze（世界边界，与S. Morey合作）。鲍里斯·伯纳斯科尼（Boris Bernaskoni）还是许多杂志的创立者和编辑，如建筑材料和A3。

an original Russian interior design exists, either shake their heads or admit that they never thought about it. How do you respond to that? Why do your professional colleagues reject the notion of a Russian interior design? It surely has to do with the term. In the Russian language there is no word for design, unlike, say, in the English language, where design is also architecture – as in the creation and design of objects. The Russian usage of the word is more superficial. Design isn't a creative act for us. Instead, it's a way to »work on the available«. It's the manipulative, form-altering way of handling objects and spaces. In the Russian language this imprecise term includes everything imaginable: everyday commodities, material, clothing. So we don't have an »urban design,« for example. Instead, we talk about city planning or city development. Seen in that context, the term design seems vague and rather limited at the same time, and only expressive when used in reference to objects or graphic products.

If you find the term design too imprecise, could you define for me space or designed space from an architectural perspective?

Do you mean space in the architectural sense or a concrete interior space? How do those differ for you? Strictly speaking, we no longer have architecture in the sense of a practice limited solely to the construction of buildings. An advanced understanding of architecture imports every discipline somehow related to art. It includes marketing, branding, economics, graphic design, and advertising. In other words, everything put to use in order to win users and future clients. Today architecture functions more like an interface between space in and of itself and the user. The user initiates the plot in the space; the architect provides the design framework, which – in my projects – is naturally determined by me alone...

Which style does your architecture follow?

If you're asking about my style, I can only answer: I don't have one. In my opinion, there are no styles anymore.

And if a client asks you?

Eclecticism! Today that term is used to degrade for the most part. In the sense of labeling uncreative a working method suspected of engaging in plagiarism. In ancient times, however, the same term had a thoroughly positive meaning: Cicero, for example; he was considered an eclectic. This referred to the philosophical method of processing what was recognized as positive from a variety of systems and having this ultimately yield a new series of teachings. Applied to

Idea Factory
BBDO Advertising Agency

architecture this would mean choosing from all the good examples the best one and then composing a new entirety from it. In any case, that's how I work with my clients. I begin by supporting them during the research phase while they search for effective solutions. Then I help them make the right decision regarding the design.

For the photo shoot, you led us into the Yeliseyevsky department store, where you posed in front of an oversized bracket console. I got the impression that you had a special affinity for this historic, almost festive architecture. What could possibly connect a young architect like yourself to architecture from the time of the Czars?

In the Russian language, we have the word »naryad « (   ). It can be used to describe, for example, festive garments. But it's also used to describe things opulent. An extravagant room, for example. »Festive« and »opulent« frequently appear in the Russian vernacular. We're not as touchy as you are in Germany when it comes to terms like these. Think of the GUM department store, the buildings in The Kremlin, the architecture from the Stalinist period – or finally the opulent, interior design of the Yeliseyevsky department store. Like a body, architecture should be clothed as well. It doesn't matter what type of fashion I choose to cover this body with. No differently than for me, as a Russian, the term »naryad« is hardly taboo, the ornate and exquisitely detailed architecture of the past is hardly taboo for me either.

否认要么承认他们从来没有考虑过这一问题。对此您怎么看？

这应该是必须面对的问题。在俄语里，没有"设计"这个词，而在英文中则不同，英文的"设计"也是"建筑"的意思，恰恰表示创作和物体设计。对我们来说，设计不是一种创作行为。在俄语里，这个不精确的词包括一切可能的事物：每天的日用品、材料、衣物，等等。所以，我们没有一个"都市设计"的概念。相反，我们谈论的是城市规划或城市发展。从那个角度看，"设计"这个词显得含糊不清，同时意思更受限制，并且只有指向物体或图片作品时才富有表现力。

如果您发现"设计"这个词太不精确，那么您能从建筑学的视角为我界定一下空间或设计空间的定义吗？

严格说来，如果从单纯限定在建筑物构建的实践层面来看，我们不再有建筑学的概念。对建筑学的先进的理解应引入与艺术有关的一切可能的学科。它包括营销、商标、经济学、图片设计和广告。今天，建筑学的功能更像是一个介于设计空间本身和用户之间的界面。用户首先提供空间场所，建筑师提供设计框架——这些事在我的项目中，自然由我单独来决定。

您所设计的建筑追随哪种风格样式？

如果您问到我的风格样式，我只能回答：我没有一个固定的风格样式。在我看来，已不再有固定的风格样式。

那如果客户要求您呢？

折衷主义！今天，那样的词汇在大多数情况下被赋予贬低意味。在意识方面，没有创造性的标记，一个工作方法会被怀疑为进行了剽窃。然而，在古代，同一个词汇却有着迥然不同的积极意义：例如，西塞罗（Cicero）被认为是一个折衷主义者。在建筑学中，这意味着从所有好的范例样本中选出最佳的一个，然后在它的基础上再组成一个全新的整体，这是我与客户配合工作的方式。在研究探索阶段，此时客户在寻求有效的解决方案，我通过帮助他们来开始自己的工作。然后，我帮助他们做出关于设计的正确决定。

为了照片拍摄，您带领我们进入了Jelissejewski百货商店，在那里，您在一个特大型的支架控制台前面摆开了架势。我印象非常深刻，您对这一历史的、近于欢乐的建筑有着特殊的亲和力和吸引力。究竟是什么使得像您这样的一个年轻建筑师与沙皇时代的建筑联系起来了呢？

在俄语里，我们有一个词"narjad"（наряд）。它可以用于描述，例如，欢乐盛装。但它也可以用于描述事物丰富的，例如，一间奢侈的屋子。"欢乐的"和"丰富的"这两个词频繁地出现于俄语白话中。像这样的词，我们并不像你们在德国那样难以处理。想一想Gum百货店、位于Kremlin的建筑物、斯大林时代的建筑，或最后的拥有丰富室内设计的Jelissejewski百货商店就知道了。就像人的躯体需要衣服一样，建筑也是如此。我选择何种类型的时尚去包裹这一躯体并不紧要。作为一个俄国人，我并没什么不同。对我来说，词"наряд"几乎不是禁忌，过去的那些华丽的、精巧的、注重细节的建筑对我来说也几乎不是禁忌。

访谈摘要：

鲍里斯·伯纳斯科尼（Boris Bernaskoni）的设计工作展示了一个非比寻常的宽阔领域。它既包括小的项目，例如家具设计和固定构建，也包括大规模的项目，像公共建筑和大型都市区域规划。然而，鲍里斯·伯纳斯科尼（Boris Bernaskoni）的所有设计都有它们自己独特的清晰而又彻底的反讽性质，这是年轻的建筑师为了将自己与主流设计区分开来。鲍里斯·伯纳斯科尼（Boris Bernaskoni）非常关注设计的细节和功能，因此，由他设计的建筑既有美好的风格样式，又有高水平的新生活构想。

您认为真的存在特定的俄罗斯室内设计吗？

是的，我这么认为。然而，我认为室内设计与合理性方面紧密关联，这就使得俄罗斯室内设计令人讶异的不合理，因为这些设计过多依赖于放纵和秀美，而不是该有的合理规则。这和那些创作此设计的人们一样，以急惰心完成工作。在德国，每个人都会提到Neufert——从1930年代以来一直为人们所接受，它代表了建筑师的最为重要的表现力。这一人类环境改造学的参考书为设计工作提供了一个清晰的框架。但在俄罗斯，这种类型的事物却从来没有出现过。在这里，每一件事情的发展更加随意和武断，更加偶然。首先，你有一个想法；然后，相信你的天性，而无需考虑细节；最后，所有问题都在于审美理念和风格样式。

根据您的想法来判断，您跟您的俄国同事没有一点相像，他们在被问到是否存在原创俄罗斯室内设计时，要么摇头

The Architect is like a cat in an empty apartment
Interview with Alexander Brodsky

我像猫一样依赖于自己的直觉

建筑师亚历山大·布罗斯基（Alexander Brodski）访谈

The work of Alexander Brodsky is characterized by a unique ability to translate traditional forms, materials, and structural engineering techniques into a contemporary language. His philosophy combines the patient search for archetypes with a marked aestheticizing of the informal. One of the most distinguishing features of Brodsky's work is the incorporation of salvaged elements from torn down buildings in totally new compositions. In this way, banal objects such as a window or a door achieve an artistic value which lends the respective project an unmistakable charm.

The number of sophisticated interior design projects in Russia is unusually high. Does this make it possible to speak today of a typically Russian interior design?

When we look at contemporary interior design in Russia, we find almost exclusively international trends. But when reviewing history, a different picture unfolds. During the Stalinist period there was an interior design only found in that form in the Soviet Union. Also during the years preceding the Revolution, and as early as the sixteenth century, we find vivid examples of a typically Russian interior design.

What are the distinguishing features of this Russian interior design?

Unmistakable is how wood is handled. The way this material was designed with during the time of the Czars is amazing. Timber architecture is characterized by a strong feeling for details and ornamentation hardly known to us today. By comparison, Stalinist-period interior design can only be described in abstract terms. A more personal feeling consumes me when I'm in one of these timber-constructed spaces. On the one hand, there is the floor plan, so unlike the floor plans of all the other residential buildings in the Soviet Union: high ceilings, decorations, ornaments, a specific type of furniture, and lighting fixtures of the highest quality; on the other, there is, of course, that incredible wallpaper, which perhaps best documents the decline of interior design in the Soviet Union: during Chrushtschow's term of office, in the 1960s, wallpapering still appeared now and then; but by

Alexander Brodsky,
born 1955 in Moscow, grew up in a family of artists. In 1968-69, he studied at the Moscow School of Art. In 1972 he entered to the *Moscow Institute of Architecture* (MARCHI). Between 1978 and 1993, in collaboration with Ilya Utkin, he drew attention to himself worldwide by participating in international competitions and exhibitions as a representative of the so-called »Paper Architecture«. Up until 2000, while concentrating on his artistic work, Brodsky produced graphic editions, sculptures, and installations. In 2000 he founded with Yaroslav Kovalchuk an architectural bureau. Their unconventional office is located in a ruinous side-wing of the state-run *Architecture Museum* (MUAR) in Moscow.

亚历山大·布罗斯基
1955年出生在莫斯科，生长在一个艺术家的家庭里。1968年初，他在莫斯科美术学院学习。1972年他转到莫斯科建筑学院（MARCHI）学习。在1978年至1993年间，他与伊莱亚·乌特金一起合作，作为所谓的"纸艺建筑师"的一名代表参加国际性的竞赛和展览为他赢得了全世界的关注。直到2000年，在其专心致志于艺术作品的同时，布罗斯基还制作了版画、雕塑和装置物。自从那时，他与雅罗斯瓦夫·卡瓦尔舒科（Jaroslaw Kowalschuk）一起合作。他们的反传统工作室设在州立建筑博物馆（MUAR）一栋破败的侧楼内。

the time Breschnjew came to office, in the 1970s, it vanished altogether.

Today wallpaper is fashionable again. Almost all the newly opened restaurants and cafés show off ornamented wallpaper. Is this the renaissance of Russian interior design?

No, not at all. These are merely examples found in international fashion magazines. To speak of a renaissance, Russian architecture as well as a specific interior design would have to be newly developed first.

But the name Alexander Brodski stands for the epitome of new Russian architecture. In any case the international architecture critics say that you have an enormous regional influence...

... it's more of a personal influence. I approach my projects on the basis of my feelings and not intellectually – and certainly not scientifically. I wouldn't like to see my work classified as belonging to a Russian tradition. When I stand in front of an empty space, I rely most on my intuition. It's like letting a cat run free through an empty apartment and then waiting for it to lie down somewhere. That's where you put the bed. The same way that a cat reacts intuitively, I rely on secret voices

in my head while I work. A case in point is the conference table I'm sitting at right now. I feel as though its placement in the space is perfect: comfortably positioned and creating a welcoming gesture toward the entrance. You only have to rely on your intuition. Like the cat in the empty apartment.

You have produced entire houses without a plan. Did that happen using the same intuition you just described? How do your clients react when they can't tell for sure in what direction your architecture is developing?

You can only work without a plan part of the time. Or in the case of special construction tasks. The restaurant you're referring to was created without a plan simply because we lacked the time. But we built a model for the client. After we presented it to him, he asked when the constructing would begin. When I requested a four-week processing period for the final planning, his answer was to the point: "We don't have the time!" So the constructing began the following day, and we used the finished model as our basis. Meanwhile, the restaurant has been standing for six years. Another project we built without a plan was the pavilion for vodka ceremonies. In that case, submitting a few rough details on the measurements was enough. The clients trust me.

95° restaurant at the
Bukhta Radosti (Bay of Joy)
2000

And you always trust yourself?

It's a lot simpler, of course, when I'm building something for myself. Every year I have my datscha rebuilt. I hire inexpensive laborers, give them a rough idea of what has to be done, and leave money for the materials. Then I leave the site and come back two weeks later. I always find it amusing to see how the workers use the last, square centimeters of chipboard so that nothing is wasted. So far I've always been satisfied with the results.

Meanwhile your drawings have made you famous worldwide. Now there is the artist Alexander Brodski, who draws fictitious spaces, and there is the architect Alexander Brodksy, who builds spaces without drawings. How do the drawings and the architecture interact?

The drawing is only a way to illustrate ideas. It's extremely important for an architect.

Everything begins here. Even convincing oneself of the idea. But I want to stress that a classic handmade drawing is always preferred over a computer-generated image. Think of how lifeless digital renderings look when you don't invest enough time in them! I see this confirmed over and over again: no sooner a project is completed, the first handmade sketches – in spite of their abstractness – always correspond with the constructed results. The proportions and the sense of space feel perfect. In order to create a usable computer-generated image, you have to establish all the project's details and materials at the start. No serious architect can do that.

Is there an ideal space for you?

I always have an ideal space in my mind. It's not a building but rather an interior space, a kind of interior court. In my dreams, I always see the same space, whether it's a living room or a restaurant. The more I can turn fragments of these visions into realities through my projects, the more interesting the results are for me. But my ideal space, mind you, is a space that I've never actually seen in reality. I'm still working on it.

Pavilion for vodka ceremonies,
Klyazminskoe Reservoir rest area
2003

访谈摘要：

亚历山大·布罗斯基作品的特点在于通过一种特殊的能力将传统的形态、材料和建筑工程技术诠释为一种当代的语言。他的哲学理念使其将对原型的耐心搜寻与普通物体的显著的审美化结合起来。布罗斯基作品中最具特色的特点之一是将拆毁建筑的废弃元素合并成为全新的组成部分。这样，普普通通的物体，如一扇窗子或者一扇门就获得了艺术价值，这也赋予了每一项工程毋庸置疑的魅力。

在俄罗斯许多久经世故的室内设计项目都是出奇的宏伟高大。我们可以说这是今天俄罗斯室内设计典型特点吗？

当我们审视俄罗斯现代室内设计时，会发现它们几乎无一例外的存在向国际化方向发展的倾向。但是当回顾历史时，一幅不同的画卷展现在我们面前。斯大林时期的一种室内设计只有在苏联的那种建筑中才能找到。同样在革命前以及16世纪早期，我们可以找到典型俄罗斯室内设计的生动范例。

这种俄罗斯室内设计的特点是什么呢？

很明显是对于木料的处理上。在沙皇时期这种材料的使用方法是非常了不起的。木质建筑的特点就是其细节处和装饰物的强烈感觉，这是我们今天很难体会得到的。比较而言，斯大林时期的室内设计仅能以抽象词汇来描述。当我置身于一个木结构的空间内，更加强烈的个性化感觉占据了我的内心。一方面，它有一个不同于苏联所有其他住宅楼的建筑平面图：高高的天花板、装潢、饰品，特殊风格的家具和最高质量的照明设施；当然另一方面还有简直令人难以置信的壁纸，这可能是记载苏联室内设计由盛到衰的最佳资料：在1960年赫鲁晓夫时期的办公室内设计中，壁纸还是偶尔可以见到的；但是到1970年勃列日涅夫时期，办公室的设计中，壁纸则完全消失了。

在没有设计图的情况下您完成了整栋房子设计。那些设计都是运用刚刚描述的同一种直觉吗？当您的客户确实不能告诉您建筑将要向哪个方向开发的时候，他们如何反应呢？

在有些时候你只能在没有平面图的情况下工作。也有一些特殊结构的建筑任务。您谈到的那间餐馆就是在没有平面图的情况下创造的，当时我需要四个星期的时间来修改完成最后的方案，客户得知后，只说了一点："我们没有时间了！"因此，这项工程第二天就开工了。这间餐馆已经经营了六年的时间。另一个我们没有使用平面图的工程是为伏特加酒盛典修建的一座亭子。在那种情况下，提交几个简略的测量细节资料就足够了。客户信任我。

对于您来说存在一个完美的空间吗？

在我的脑海里总是存在一个完美的空间。它不是一栋建筑，更不是一个室内空间或者一种内部庭院。在我的梦中，我总是看见同一个空间，它是一间起居室也可能是一间餐馆。我能将更多想象中的碎片转化进入我的现实项目中，就我来说结果是我更感兴趣的。但是我的完美空间，你提到的，实际上是一个在现实中我从未见过的空间。我仍然一直在努力的实现它。

Clients used to be more curious

Interview with Alexey Kozyr

"20世纪90年代的客户对于建筑的好奇心远远大于今日。"

建筑师艾力克赛·科西尔（Alexei Kosir）访谈

Alexey Kozyr is among the representatives of Russian High-tech architecture. His romantic interpretation of this style led to his enjoying international attention over the last few years. His unique handling of metal, glass, and light transforms banal built-in structures into extravagant installations styled to the last screw. The combining of design and engineering arts as well as structural thinking and high-quality craftsmanship lend his interior design an almost museum-like quality. Bizarre exhibition pieces take on everyday functions.

Is there a specifically Russian interior design? As someone so well-known in public for his designs of interior spaces, I'm sure you've given this question some thought.

First we have to realize that buildings are being constructed and utilized. So there is, of course, interior design. Given the number of newspapers and magazines devoted to this theme, and the number of businesses that specialize in producing apartment furnishings, we can surely speak of interior design in Russia. There's a market for it. However, this is merely a socio-cultural finding. If you ask for my opinion as an architect, I would give you a different answer. I would say that 80 per cent of everything built here over the last few years was only a copy of what existed in the West. Russia has no high-quality interior design tradition. It's possible that everything was wiped out by seventy years of the Soviet Union. Yes, at present things are definitely developing here. But it's hard for me to say whether this concerns something "genuinely Russian" or simply imported merchandise from the West.

You studied in London. How did the time spent at the Architectural Association in London influence your work in Russia?

Intellectually, my life was greatly altered there. At least my approaches. Unfortunately, I was in Great Britain only a year, hardly long enough to really delve into its origins and traditions. When I returned to Russia, I found it almost impossible to apply the know-how acquired in London.

Alexey Kozyr
completed his architecture studies in 1993 in Moscow, followed by a year at the *Architectural Association* (AA) in London. In 1994, in collaboration with Natalya Lobanova and Ivan Chuvelev, he founded the office ARCH 4, soon to become famous in the field of interior design. The projects completed here not only included furnishings for public buildings and private clients but also classic Russian country houses. The author of numerous publications focussed on interior design, he works as the curator of exhibitions as well. Since 2006 Kozyr manages his own office, a work space situated on ulitsa Tverskaya in Moscow, in an apartment recently purchased from one of Stalin's grandchildren.

艾力克赛•科西尔
1993年完成了他在莫斯科建筑学的学习生涯，之后他在伦敦建筑协会（AA）工作了一年。1994年与N. Lobanowa和I. Chuwelew共同合作创办了ARCH 4工作室，之后不久就在室内设计领域名声大噪。所承接的项目不仅包括公共建筑和私人客户的室内陈设的设计，也包括俄罗斯经典乡村房屋的设计。由于已经出版了许多关于室内设计这一主题的出版物，所以他也担任展览的策展人。自从2006年，科西尔开始管理自己的工作室，这间工作室位于Twerskaja街一套最近刚从斯大林的孙子手中购买的公寓里。

Although there was a professional demand in the form of clients, the clients' interest in innovation lagged behind my expectations. There was one exception: the client of Airplane Apartments. He handed me a blank sheet of paper and gave me the freedom to do as I pleased.

You were in London over ten years ago. Now you're describing problems you had after returning to Russia. How do today's clients differ from the ones who took advantage of your services in the mid-1990s?

In the 1990s things were livelier, crazier. You noticed a kind of adventurer's spirit among the clients then; they were curious about architecture. A great deal has changed. Today everyone thinks they know everything. They would never listen to an architect! In the 1990s, the building of a new apartment qualified as a cultural event. It was something looked on with admiration. Today everything is routine and static. In the past even the way that architects handled new tasks was different. It was normal to give all you had – 200 per cent of yourself – to a project. It was more than just a matter of working to earn money.

You say the 1990s were exciting. So how do you motivate yourself today regarding new projects?

You mentioned that, in the 1990s, people were more interested in impulses from abroad, and that they weren't afraid of experimentation. Then you said that today everything is less surprising. But isn't it sometimes advantageous for an architect to work with a client who knows exactly what he or she wants?

Those are two entirely different questions. Regarding the clients, I prefer a situation like the one already described, where I'm given a blank sheet of paper. What I frequently experience nowadays is that a client comes to me with a page ripped out of a magazine and asks me to build a clone of the depicted building. One's own ideas are no longer in demand. Regarding the other question, about motivation, for many architects the creative process is focussed more on furniture design in the meantime. In my case, I devote more and more time to working artistically – to creating installations and related pieces.

In conjunction with your work as an interior designer, do you see the artistic work as an independent variable?

There's hardly an architect who can really separate these two areas. They correspond in too many ways. There was

33

NEW INTERIOR DESIGN IN RUSSIA

Apartment on Tishinsky
1999

House in Gorki
1999

an apartment house, for example, and a staircase had to be designed for it. In the end, the staircase resembled a concert piano except that it didn't produce tones. I'm only referring to the close connection between architecture and art here.

How do you go about your work? What happens after a client calls you? Do you sit down at a computer or reach for a blank sheet of paper? What happens?

First we meet, the client and myself, and we have a long conversation. We discuss more the actual project; we discuss everything imaginable. While we talk I try to sense what the client wants and to what extent the client is willing to trust me. We don't study books or other materials together. Later I try a few trial designs, ideas developed from our talk, from what the client likes and doesn't like. I don't use drawings or study models during this process.

How important is architectural theory for your work?

Even if I don't think of myself as a theorist, I realize the effects of theory are everywhere. Yet I never encase my design in a theoretical-abstract superstructure. My sense of theory unfolds within the product itself, within what I've actually created. For me, theory is what I build.

Now that we discussed space and theory, I wanted to mention the exhibition presented under the heading "Lust auf Raum" (In the Mood for Space). What's the first thing that makes you think of? What do you expect?

I would imagine a dark space, haunted here and there by light. An endless space. Although I cross it, I never reach my destination.

VIP floor »Roof«
on ulitsa Rochdelskaya
2001

访谈摘要：

艾力克赛·科西尔是俄罗斯高科技建筑领域的领军人物。在过去的几年里他对于这种风格的传奇阐释引起了国际关注和好评。他对于金属、玻璃和光线的独特处理将普通的内置结构转变成为奢华的固定装置，设计无处不在乃至最后一颗螺丝钉。设计与土木工程艺术和构造观念以及高科技技术的结合赋予了他的设计几乎如同博物馆一样的品质。奇异的展品行使着每日的职能。

俄罗斯室内设计有特定的样式吗？作为一位在室内空间设计领域如此闻名的设计师，我相信您对这个问题一定也有一些思考。

首先我们必须认识到建筑是被建造并且使用的。因此当然室内的设计是必需的。从大量开办的专门致力于报道这一领域的报纸和杂志以及数量可观的专门生产公寓配套家具商家来说，我们可以肯定地说在俄罗斯有室内装饰业。因为它拥有广阔的市场。但是，这仅仅是一个社会文化的发现物。如果你问我作为一名建筑师的观点，我愿意给你一个不同的答案。我会说在过去的几年里这里百分之八十的建筑仅仅是西方已有建筑的复制品。俄罗斯没有高科技室内设计的传统。在苏联的70年里百废待兴，这种情况是存在的。是的，目前这里肯定是在发展。但是我很难说这种发展是"名副其实的俄罗斯式风格"抑或仅仅又是西方商品的简单输入呢？

您谈到90年代是充满刺激的年代。那么在今天当您面对新项目的时候是如何激励自己的呢？您提到在90年代人们更

感兴趣于来自国外的设计冲动，并且他们不害怕成为试验品。然后您谈到今天的每件事情都缺少了新鲜感。但是有时候一位建筑师与一位明确地知道自己想要什么的客户合作难道不是更有利于工作吗？

那是两个完全不同的问题。关于客户，我宁愿是我上述所提及的那样的客户，交给我一页空白纸。现在我经常会遇到这样的客人，他拿给我从杂志中撕下的一页纸，让我依据上面描述的建筑克隆一份。建筑师自己的创意不再被需要。另一个问题，关于激励，对于许多建筑师来说创造过程中将更多的精力集中于家具的设计上。在我的设计案例中，我则花费越来越多的时间用于艺术方面的考量——去创造空间的设置以及相关空间的和谐处理。

建筑理论对于您的工作起到了怎样重要的作用呢？

尽管我不认为自己是一位理论家，但是我认识到理论的影响无处不在。然而我从来都不会把我的设计封闭在一个抽象理论的上层建筑中。我理解的理论是呈现在产品自身内，在我实际已经创造的作品中。对于我，理论就是我建造的项目。

现在我们讨论空间和理论的问题，我想要说的是标题为"情绪空间"的展览会。您首要考虑的问题是什么？您期望的又是什么呢？

我想象的是一个黑暗的空间，四处萦绕着灯光形成的鬼魅空间。那是一个没有尽头的空间。尽管我在穿越它，但是我将从来都不会到达我的目的地。

Spectacular architecture is of short life

Interview with Mikhail Filippov

"壮观建筑的短命是预先注定的。"

建筑师米哈伊尔•菲利普瓦（Michail Filippow）访谈

One glimpse at his drawings is enough. Uniquely captured here, viewers recognize the Old Masters' approach to architecture through artfully executed sketches and perspectives. They reflect the classical luxury of the Czar times. In his work, Filippov disclaims the influence of Russian Constructivism regarding it as a great damage to the world culture similar to the impact of the October revolution to social life. In his work, Filippov tries to revive Russian neoclassical tradition of the Silver Age lost in the period of Constructivism.

In your opinion is there a specifically Russian interior design?

I can only answer that question with difficulty. For one thing, I don't particularly like the term 'design,' no matter how one uses it – whether in the Russian, European, or American context. I don't understand what it's supposed to mean and I have no feeling for it.

The term 'design' or the design object?

I really am involved with differentiating between so-called design objects and interior design. For me, the latter has less to do with the commonly used design term and much more with architecture. Then comes the question of how to deal with space, how to form and define it. The problem with contemporary architecture is

that it sees space as a design object. Consider the age of Palladio, when objects found in rooms and buildings had no stylistic connection to architecture. There was a clear difference between the objects and their surroundings then. Whereas today the same aesthetic parameters are applied to all design objects, whether the project concerns a sofa or a skyscraper. It's all a question of scale now, and the stylistic treatment of the object no longer varies. This type of conceptually constant treatment satisfies a variety of tasks. An object can be enlarged or reduced. What exists as a door handle one minute can become a couch or a building the next. I find that suspicious.

You represent an interesting position. Today it's hard to find an architect who speaks about a dominating trend or overall tendency in the

design discourse. How does this appear to you?

Well, design exists. There's no denying that. Yet there is no longer architecture. At least not when it functions by using the same principle as design. Design means for me today what it always meant: the drafting and designing of moveable objects. Architecture is immobile. That's the difference. These moveable, dynamic, design objects always refer to the fixed structures provided by architecture. Modern architecture changed all that. It all began during the first decade of the twentieth century when concrete came into play and architects started to design buildings as if they were moveable objects. The whole of contemporary architecture is characterized by this principle of instability. And conceptual sameness is no longer rooted in architectural principles. It emerges directly from the design object. But if you remain loyal to the classic difference between architecture and object design, you also accept their respective, specific principles.

Today they say "Anything goes". Has architecture lost its laws?

Maybe architecture has relinquished its laws but not its nature, not its essence. It's possible that no one has ever demonstrated that as convincingly as Le Corbusier. What

he created was hardly a joke but rather a far-reaching confrontation with the gestures and the essence of architecture. Whether New Style or Old style, architecture is architecture.

The New Style is always a further development of what already exists. This premise can be traced back to ancient times: the New has always been intrinsically related to the Old. The same applied to the Renaissance with its emphasis on horizontals, and to the vertical-loving Baroque period that followed it: these were always confrontations with tectonics, or with the tectonic rules of architecture. Modern architecture with its star cult has either forgotten these rules or decided to ignore them. How could it have become so popular to simply throw these traditions overboard and then try to repeatedly reinvent architecture?

In my opinion, none of these modern 'Pop-star' architects have discovered anything new. These architects merely represent the classical modern tradition of the previous century's 1920s and 1930s, since all of modern architecture evolves from this paradigm. Even the postmodernists offer little in the way of innovation. In many respects they only instigated a return to what already existed. This is a revival

Palazzo Grimani
paper, pencil

architecture developed from an illusory concept of freedom. Before the dawn of modernism there existed no demand in architecture for any formal uniqueness, for creative innovation on a grand scale, or anything else along those lines. Architecture oriented itself toward the past, toward something possibly a fantasy-ridden "golden age". Now the embracing of such tradition no longer takes place, because the architects who determine the rules today are more so designers, and they obey the rules of the market. A client of mine, a food concern manager familiar with all the tricks of the trade in the market economy, says: "Money dictates architecture."

Can we stay with that example? Why do you think Perrault's design could be so successful while competing against the work of tradition-oriented participants?

In conjunction with this competition, an architect from St. Petersburg demanded of the jury that, during the decision-making process, its members consider the respective planner's familiarity with the task's urban development as well as the architectural context, and therefore include an architect who came from the city and knew it well. Precisely this aspect was a definite committee criteria. And the architect was even excluded

from the competition.

Then why did the City of St. Petersburg invite international architects who clearly intended to produce a design object on the premises? In addition, this was obviously related in some way to the competitors' monetary awards and the commissioning parties or clients...

St. Petersburg is not only an exception in Russia, but also when viewed in the global context. Conservation orders function rigidly here, and whenever a new structure is built people expect something spectacular enough to create an exciting contrast to what already exists. We're talking about a precedence-setting architecture, which consciously breaks the rules and laws of the historical. Take Gazprom's crazy skyscraper. It's being built. This has nothing to do with the building or its purpose. It's all about creating a provocation. This controversial development only applies to our generation. But the short life span of spectacular architecture is, so to speak, preprogrammed. Sooner or later these buildings are torn down again, a current example being the "Rossija" Hotel in Moscow.

San Trovaso
paper, water-colour

访谈摘要：

他的设计图只需惊鸿一瞥就足够了。独独吸引在此，观看者认出那些通过巧妙手法制成的草图和透视图是早期绘画大师对于建筑的把握，同时也体现着沙皇时代巴洛克风格的盛况以及继承了塔特林一样的前辈创建的俄罗斯构成主义的丰富遗产。通过其作品，菲利普瓦探索了这一动人的领域，这也是一个由历史的伟人居住的巨大的文化领域。他鲜明的古典风格表明了一种宽宏的姿态，这显示了对伟大传统的尊敬。

以您的观点，您认为俄罗斯室内设计有一个特定的样式吗？

我只能回答说这个问题很困难。首先，我特别不喜欢这个词"设计"，不管谁如何使用它——无论在俄罗斯、欧洲还是在美国。我不理解这个词被假设的意思，并且我对它没有感觉。

这个词 "设计"或者称为设计物呢？

我确实非常困惑于区分所谓的设计物和室内设计。对于我来说，后者在一般性的设计中运用的较少些，而更多的应用于建筑。然后是如何处理空间，如何构成和定义空间的问题。当代建筑存在的问题是把空间视为设计物。回顾帕拉蒂奥时代，当时在房间和建筑物内找到的物体与建筑之间并没有风格上的联系。那时在物体和它们的环境之间存在着一个非常清晰的差异。但是今天同一个美学元素被应用在全部的设计物上，无论项目是关于一张沙发还是一幢摩天大楼。现在就是个比例问题，并且物体的风格处理不再多样化。这种处理风格满足了多种设计任务。

您代表了一个有趣的立场。今天很难找出一位建筑师在设计演讲中谈论室内设计的主要趋势或者全面的倾向。您如何看待这一问题呢？

好，设计是存在的。没有人否定那一点。然而不再有建筑。至少当建筑通过使用与设计相同的原则行使它的功能时，它就不存在了。就我来说今天的设计意味着其本意：对于可移动物体的设计。建筑是稳定的，这就是不同之处。那些可移动的、动态的设计物总是与建筑本身固定的结构联系在一起。现代的建筑改变了这一切。

我们能与那个例子与时俱进吗？为什么您认为佩罗的设计在与主张传统导向的参赛者作品竞争时会取得成功呢？

这场竞争，在决策过程中一位来自圣彼得堡的建筑师要求评判委员会成员考虑各个设计者对于这个项目的城市发展以及建筑环境的熟悉程度，因此委员会中包含了一位来自这个城市并且熟知这一切的建筑师。恰好这一点是委员会标准中一个明确的规定。并且这位建筑师甚至一度被排除在这场竞争之外。可能这一短寿的情况正好是该建筑展示给我们的它们的类似物体的特性。如这些建筑一样，它们的结构具有完全类似物体的特性，到处可见，很少关心特殊的环境，更少考虑城市的规划或者建筑的分布状态。因此对于这一点还是有利的：当这些建筑被拆毁的时候不会直接伤害周围的环境，这一点着重强调的是该建筑的恣意妄为。

Shop »Arsenal« 军械库
Architects' Office 建筑师工作室
Office of a Direction Company 主管办公室

ABD Architects ABD 建筑师事务所

Shop »Arsenal«
军械商店

建筑师 B. Levyant, N. Sidorova, D. Loren
位置 Moscow, prospekt Mira 26, building 1
面积 800 m²
完工日期 2004

类别 office
摄影师 A. Rusov

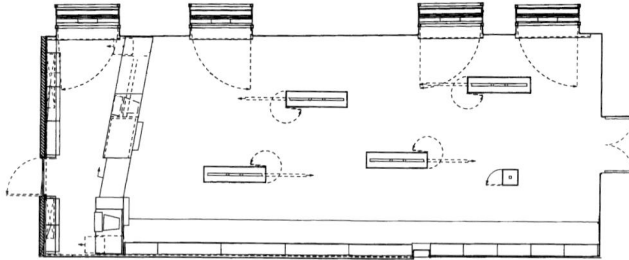

floor plan
show room

平面图
展示间

The shop's interior design expresses the dual character which weapons have in modern society. In certain contexts they are treated as quite normal goods for sale, on the other they are shown as museum exhibits. And it in this dual aspect that they are presented here. Largely isolated from the outer word – an isolation intended by its architectural design – the shop generates a museum-like, culturally excessive atmosphere. Neither light nor noise from the street penetrate the curtained windows. The highly polished granite with which floor and walls are covered symbolises exclusivity and wealth. For weapons are a luxury. Their cold, smooth surfaces give the whole room an abstract appearance. The rifles, pistols and revolvers rely totally upon themselves for their effect on the observer – and upon the light, which dramatically strengthens this powerful self awareness. The highly concentrated precision with which the interior of this shop has been designed forestalls the potentially frightening aspects of the items on display.

店铺内部设计充分表达了武器在现代社会中的双重特性。在某种情况下它们被作为正常的货物用来销售，在另一方面它们如同博物馆里的展品一样用于展览。这家店基于它的这一双重性存在于此。与外界很大程度的隔离——是有意通过建筑设计造成的一种隔离——使这家店产生了一种类似于博物馆的过度的文化氛围。街道的灯光和噪音都不能穿透窗帘密闭的窗户。覆盖着巨大的光洁大理石地面和墙壁更表明它的排他性和财富。因为武器是奢侈品。它们冰冷、光滑的表面为整个房间带来抽象化的外表。步枪、手枪和左轮手枪完全依赖着自身影响着观察者——在光的作用下显著增强了这种强大的自我意识。店铺内部高度集中的精确性设计有效遏制了展品潜在的令人生畏的一面。

Architects' Office
建筑师工作室

建筑师 B. Levyant, M. Hellavi, P. Policarpov
建筑公司 ABD Architects
位置 Moscow, prospekt Mira 26, building 1
面积 800 m^2
完工日期 2004

类别 office
摄影师 A. Rusov

third floor plan
第三层平面图

If the rooms in this architects' office reminds you of a luxurious apartment in New York – well, that is their intention. For the favoured location of the building by the Botanical Gardens and the sheer size of the space available definitely have something of the qualities of a premium location. The interior stretches across the two upper storeys of a four-storey modern building. The complex ground plan has been translated into a difficult, technology-oriented design. The reception area leads to a fully-glassed lift, which leads to the work rooms. At the other end is a staff canteen, supplied by a service lift from the restaurant below. The architectural and design departments are separate, but they use the library, archives and design studio jointly. The project managers sit in imposing offices, separated by walls of matt glass from everything happening outside. An open staircase leads to the upper storey.

如果建筑师工作室的房间让你联想到了纽约奢华的公寓——好，那就是他们最想达到的目的。植物园附近优越的位置和可用空间绝对的面积确实有几分优质地段的味道。内部空间占据了这一现代四层建筑的三四层。复杂的平面图已经被诠释成为一种难度很大的技术型设计。接待区通向一部玻璃电梯处，从那里可以通向工作区。在另一端是员工餐厅，由楼下的餐馆通过一部工作电梯配送饭食。建筑和设计部门是分开的，但是他们共同使用图书馆、档案和设计工作室。工程的主管们坐在富丽堂皇的办公室里，一堵毛面玻璃墙将其与喧嚣的外界隔开。一部宽敞的楼梯通向上一层。

stairs to the 3th floor along
the glazed wall facing the
Botanical Gardens
通往三楼的楼梯
面朝Botanical花园的釉面墙

47

reception
working room

接待处
工作间

meeting room area
upper floor reception
counter; behind the mat
glass partition there is a
negotiation room and offices

会议室接待区域

Director's Office
经理办公室

建筑师 A. Zhidkov
位置 Moscow
面积 900 m²
完工日期 2005

类别 office
摄影师 A. Rusov

floor plan
main corridor
平面图
主要通道

This spacious office in a Moscow business centre has been divided into three areas, on the client's instructions. Along with rooms for dealing with the public, and rooms for internal communication, there is the normal working area, plus offices for senior executives which, with their own reception rooms and an adjoining recreation area, take up most of the total area. The architects have paid particular attention to the entrance area. The black polished reception desk is strictly orthogonal and mounted in genuine mahogany. While the design of this area is intended to express the firm's corporate image, the director's office, and that of his deputy, suggest quite a different atmosphere: walls and furniture of warm walnut create a comfortable feeling. In the workrooms and meeting rooms, however, an austere, sober appearance is dominant. Here the primary consideration was functionality.

位于莫斯科商业中心处的这间宽敞的办公室根据其客户的要求被分成了三个区域。公共事务区、内部交流区、还有日常的工作区以及占据了绝大部分空间的高级执行官办公室，附带独立的接待室和一个相邻的娱乐室。设计师特别注意了入口处的设计。供接待客人使用的矩形的黑色桌子由红木制作而成。该区域的设计计划用来展示公司的整体形象，而经理以及其副手的办公室则是一个完全不同的氛围：墙壁和暖色调的胡桃木家具营造了舒适的感觉。然而在工作室和会议室内则主要体现严肃整洁、一丝不苟的设计。这里首要考虑的是功能性。

51

The reception area is treated as
a separate volume.

被单独划分的休闲区

53

office of the head of the
company
meeting room

公司高层会议室

NEW INTERIOR DESIGN IN RUSSIA

Apartment on Ulitsa Spiridonovka
Astradamskaya Apartment
Nikolinaya Gora House

Spiridonovka 公寓
Astradamskaya 公寓
Nikolina Gora 建筑

irapetov Levon Airapetov 建筑师事务所

Apartment on Ulitsa Spiridonovka
Spiridonovka公寓

建筑师 L. Airapetov, K. Avakov
位置 Moscow, ulitsa Spiridonovka
面积 230 m²
完工日期 1998

类别 apartment
摄影师 Z. Razutdinov

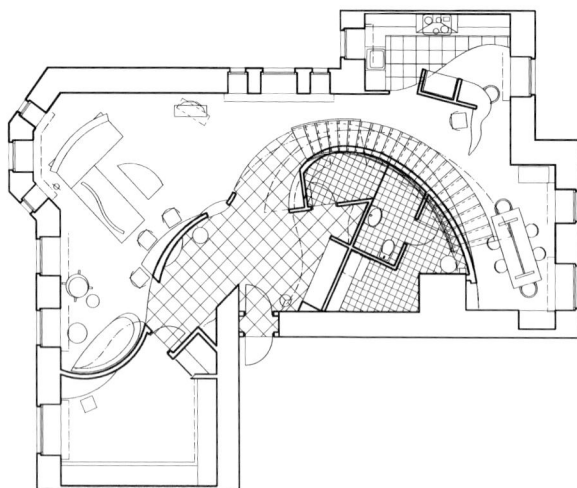

56

floor plan, lower level
living room
平面图
起居室

To put it without any frills, you could say that rooms are just a subsidiary aspect for this architect; they are derived merely from the dynamic composition of surfaces and bodies between walls. This type of approach is an obvious one when it comes to this spacious apartment in ulitsa Spiridonovka, which takes up two storeys of an older building. Alongside the reconstruction of the rooms, it seems as if a sort of internal chaos had brought the strict symmetry of the building's historic structures to an implosion. The deconstructive impetus of the design is sharpened further by the use of artificial light. The observer is led to look into depths where there are none, changing the silhouette of perceived objects. This design confronts the observer with a different sort of perception of spatial qualities, allowing him to view the room in its topographic essence. The apartment is no longer an associated context of different rooms; it must be viewed as a unit covering different locations and different passages.

房间的安排不带任何的虚饰，它们只是由墙面和墙体相互动态组合而成。对此您可能会说房间只是该建筑师的次要的作品。但是对于Malenkovskaya街上的一处宽敞的公寓来说，这种方式就显得很常见了。这处公寓占据一栋老式建筑的两层，关于房间的改造，似乎是一种内在的混乱导致建筑本身的古老结构所具有的严谨对称性发生向内的爆炸。人造光源的使用，使得设计的解构性推动力进一步尖锐化。随着可感知物体轮廓的变化，观察者被引领着探寻了一种从未有人到达过的深度。这种设计允许旁观者依照这个房间的地形学本质来观察它，所以使他们和一种对空间本质的不同理解有了面对面的接触。这间公寓不再是一个不同房间的结合体，而是被认为是用来占据不同位置和通道的单位。这些空间不是作为房间而存在的，甚至通道也从未成为楼梯或者走廊。因为不可能知道这些房间的分界线，所有在那里生活的人能做的只是在这些自由流动的形状中移动。

LEVON AIRAPETOV

living room
wall between the hall
and the living room
bathroom

起居室
大厅和起居室之间的墙壁
浴室

58

LEVON AIRAPETOV

Astradamskaya Apartment
Astradamskaya公寓

建筑师 L. Airapetov, D. Grekova
位置 Moscow, ulitsa Astradamskaya
面积 138 m^2
完工日期 2006

类别 apartment
摄影师 Z. Razutdinov

floor plan
kitchen: dining table
and bar

平面图
厨房：餐桌
和酒吧间

Actually this project was an almost everyday commission. The job was to design the apartment for a perfectly normal family of father and mother and two children. There were enough rooms – living room, bedroom, two children's rooms, bathroom and study – but they were not all that spacious. There was one great attraction for the architect, however – the almost free ground plan. The only items which planning needed to take into account were two reinforced-steel girders and the shafts for the building technology. The design follows an almost mathematical logic in its consistency. All details are derived from the spatial structure and their realisation reflects the basic principle of order which aims to unite opposites: horizontals and verticals, openness and seclusiveness, dynamism and stasis. The materials used also reflect this approach: here the contrasts are of a haptic kind, alternating between soft and hard, cosy and cool. It is a fine balance between community and solitary retirement.

为拥有一对父母和两个孩子的标准家庭设计一套公寓的工程几乎每天都会碰到。要有足够的房间，包括起居室、卧室、两间孩子的房间、浴室和书房，但是它们都不用太大。然而对于建筑师来说巨大的吸引力在于几乎可以任意发挥的平面图。平面图只需要考虑的是两个加固的钢梁和建筑技术中出现的支柱。设计的连贯性遵循着几乎精确的数理逻辑。全部细节设计都是依据空间的结构，设计师设计结果反映对立统一的基本原理：水平和垂直；公开性和私隐性、运动和静止。使用的材料也反应了这一方式：这里的比较是与触觉相关的，软和硬、舒适和凉爽之间的互化。最后，但并非最不重要，可用空间的安排表明了家庭生活耐人寻味的两级。很好的平衡了大家庭和独隐者之间关系。

Nikolinaya Gora House
Nikolina Gora公寓

建筑师 L. Airapetov, P. Romanov, A. Panchenko, V. Preobrazhenskaya
位置 Moscow, Nikolinaya gora
面积 780 m²
完工日期 2006

类别 apartment
摄影师 Z. Razutdinov

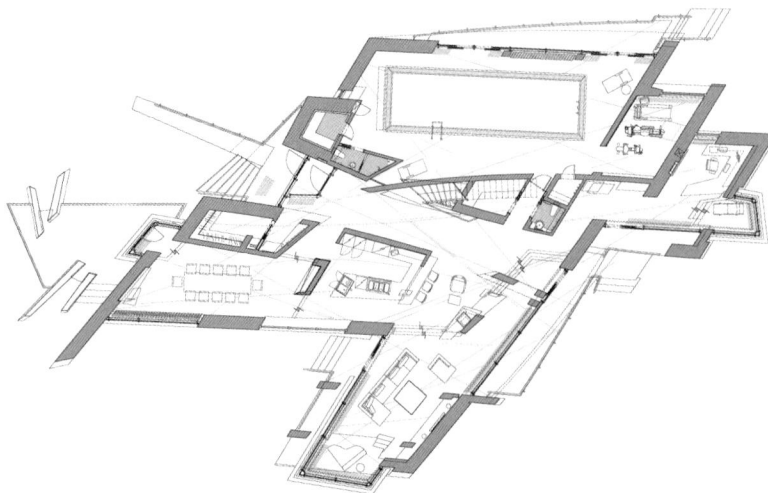

plan
entrance view of the central
block with staircase
平面图
入口处景观

Only one person lives here, but he has a lot of guests. So it was necessary to combine prestigious features with the need for a private sphere. The first floor is very open and inviting in its design and has been structured in a completely classical way, following a particular programme. The facilities here have a quasi social, open character: a spacious lobby, kitchen with table to eat on, swimming pool with fitness studio. The essential feature of this floor is transparency. Large glass frontages provide views in and out and let in plenty of daylight. The second floor has been divided into two units, of which one includes two guest rooms and a bathroom, while the other is the place where the owner can withdraw to find privacy. A glass passage leading to the roof terrace links these two units. As in other projects by this architect, the dominant colour is white, broken only by a few elegant contrasts – the wooden surrounds in the central area and the blue of the water in the swimming pool.

这所公寓只有主人一个人住在这里，但是他要在这里招待很多宾朋。因此对于该公寓必须要把久负盛名的特点与私隐空间的需求两者之间结合起来。依据独特的程序，一层的设计是全开放性的，设计也独具吸引力并且以非常经典的方式建造。这里的设施类似公众社交场合的特点：宽阔的休息室，配有餐桌的厨房，带有健身室的游泳池。一层最主要的特色是通透性。房子正面的大型玻璃可一览室内外景色，并且也可以让更多的阳光进入室内。二层被分割成两个单元，一个单元包括两间客房和一间浴室，而另一单元则是主人可以抽身独处的私人空间。以上两个单元由一条通往屋顶阳台的玻璃通道连接着。就像该设计师的其他作品一样，这所公寓主要使用白色作为主色调，偶尔会有几处雅致的对比色——中心区的木质边缘装饰和游泳池中蓝色的池水。设计师特有的设计风格营造了这里的平静祥和的氛围，尽管有时候也会有某个问题出现：这栋建筑是一处水平设计结构还是一种立方结构呢？

LEVON AIRAPETOV

staircase
ground floor, detail staircase
first floor, interior of the central distribution block
swimming pool

楼梯
地下一层，楼梯细部
一层，室内中央循环
游泳池

Airplane Apartment
VIP Floor
Lukov Pereulok Apartment
Apartment in Pushkarev Pereulok

飞机公寓
VIP 区
Lukov Pereulok 公寓
Pereulok 公寓

ARCH-4　ARCH-4 建筑师事务所

Airplane Apartment
飞机公寓

建筑师 A. Kozyr, I. Chuvelev, N. Lobanova
结构工程师 MNIITEP
钢结构设计师 I. Silverstov
施工 Bioinjector, DOT-MC
位置 Moscow, Lefortovo
面积 42 m^2
完工日期 1997

类别 apartment
摄影师 D. Livshiz

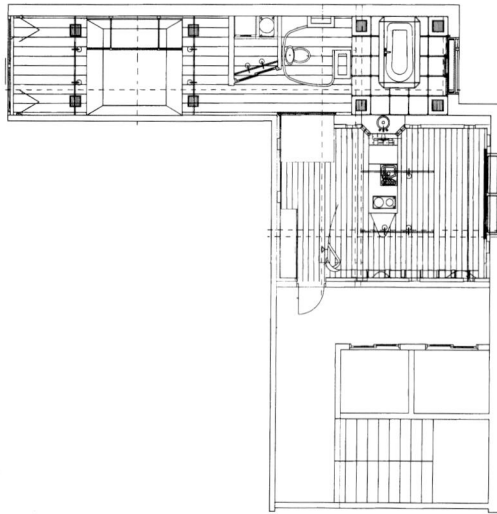

floor plan
living room, entrance area
平面图
起居室，入口

The client had good, though sentimental reasons for his choice. He decided on a modest two-room apartment in an unassuming Moscow apartment block, because he had lived there as a child. Following the conversion, of course, nothing remains to suggest this time. The conversion has created an interior which unites the remnants of the past machine age with the high-tech toys of the digital era. It could be a bit kitsch too. Aircraft parts serve as room partitions, the bath is a large goldfish aquarium containing an extra basin for the bather, and the head of the bed merges into a walkway leading onto the leaf-covered balcony. That money was no object when it came to the interior design was plain from the start. The grey fitted kitchen with the red counters was specially made in Austria. Components of old aircraft can be found all over the place, now only recognisable by their labels. Where it now says »Don't pull handle« you will find the TV.

尽管委托人对于自己的选择有些伤感，但是他确实从中受益。他做出决定的这套面积适中的两室公寓位于莫斯科一处并不招摇公寓楼内，当他还是孩子的时候就住在这里。当然随着不断的改造，没有留下多少与那个年代有关的痕迹。这一改造已经创造了一个将逝去的机器时代残迹与数码时代高科技装饰物相融合的室内布局。可能也有点庸俗。飞行器的零件成为了房间的隔墙，浴室是一个大型的金鱼缸和一个为入浴者额外准备的面盆组成的，床头与通往落叶覆盖的阳台的一条过道融为一处。从最初设计的时候就决定以简朴为主，所以资金在这里不是问题。灰白色的厨房是在澳大利亚特别定制的，它带有一个红色的吧台。旧航行器零部件随处可见，现在只能通过它们的标签来辨认。标签上说"不要下拉手柄"的地方，你将找到电视机。

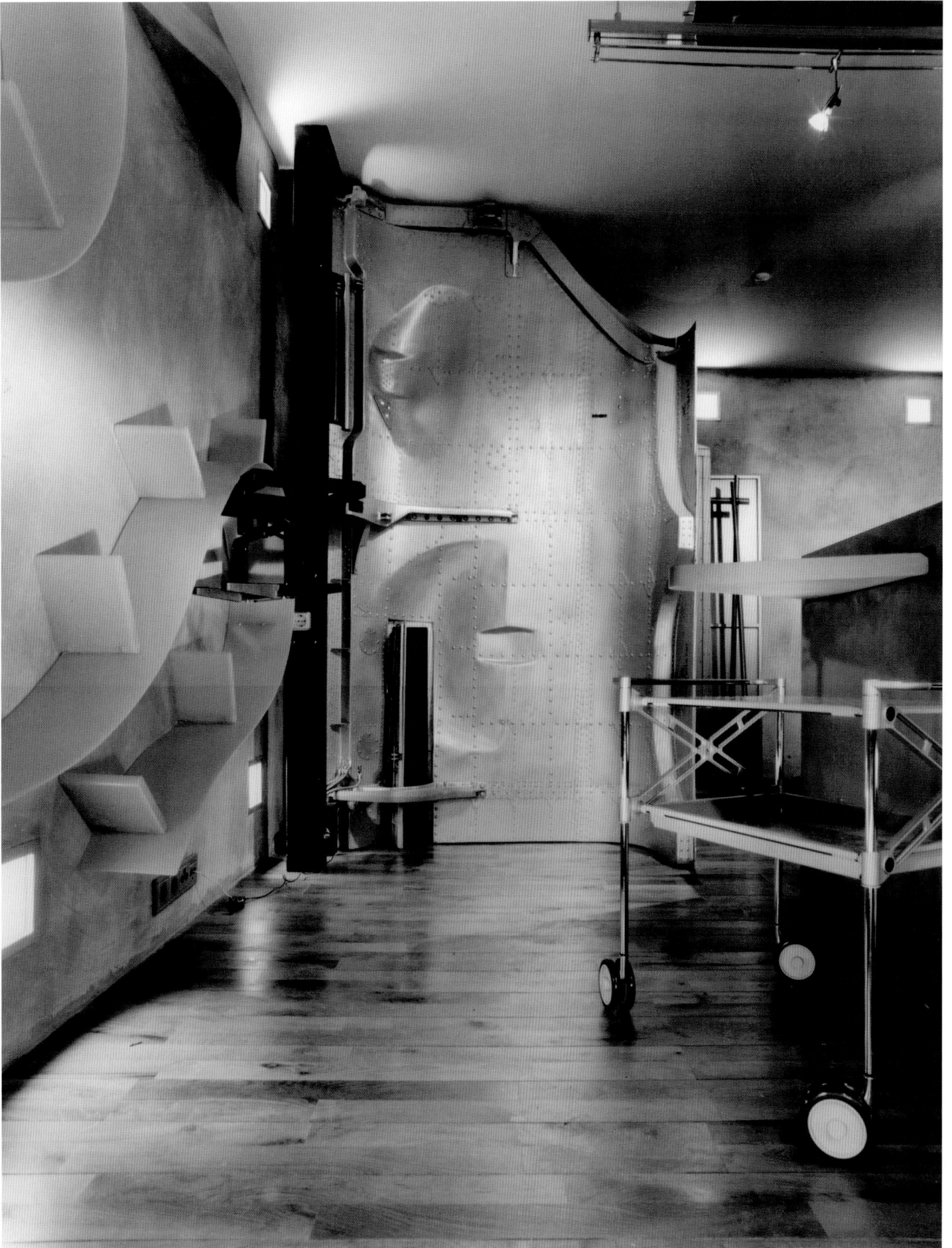

71

bathroom
with aquarium
带有养鱼池的浴室

带有养鱼池的浴室

ramp leading from the living
room to the private area
bedroom
bathroom sink
从起居室到私人区的过渡带
卧室
浴室水池

VIP Floor
VIP区

建筑师 A. Kozyr
结构工程师 A. Plotnikov
钢结构设计师 I. Silverstov
建筑公司 G. Bakitko, B. Kuznetsov
位置 Moscow, ulitsa Rochdelskaya
面积 400 m²
完工日期 2001

类别 apartment
摄影师 I. Kaidalina

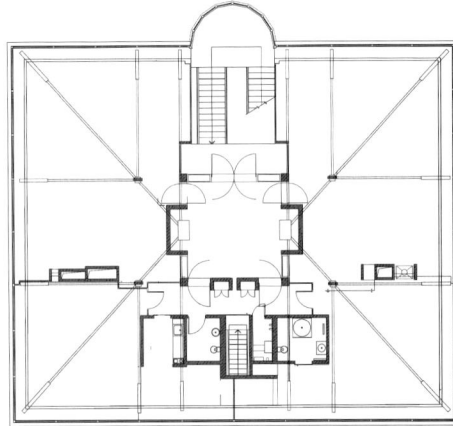

floor plan
cabinet
平面图
壁橱

The spacious, bright atmosphere is the result of a considerable structural operation. Because the spacious office floor was supplied only insufficiently with daylight, the architect simply decided to lift the roof, thus creating a light-flooded attic floor. The premises include reception area, work-rooms, conference room, and a small rest room. The choice of materials played a major role in the design. The dominant material is black-coloured concrete, now well known as the architect's par-ticular trademark. The brutal aura of this material is counterbalanced by soothing colours and materials. Numerous coats of latex paint have given the walls a silk-like sur-face; walls and floors have been covered with warm, dark woods. Only the prestigious reception room has a floor of highly polished marble. Detailed finishings in steel and glass complement the ele-gant, spacious interior.

宽敞、明亮的氛围是由一处值得考量的建筑营造出来的效果。宽阔的办公室区只凭借日光的照射是不够的，因此建筑师决定提高屋顶，结果顶楼层被淹没在阳光之中。这栋建筑包括接待区、工作室、会议室和一个小型的休息室。选用的材料在设计中发挥着主要的作用。主体材料使用黑色的混凝土，现在这已经成为这位建筑师独特的标志。具有镇静作用的色彩和材料的使用弥补了混凝土本身具有的野性特质。多次涂刷乳胶漆使得墙面如丝绸般光滑；墙和地板用温暖的暗色实木铺设。只有重要的接待处使用的是高度抛光的大理石地面。钢和玻璃结构的家具使得室内空间更加显得优雅和宽敞。

corridor

走廊

meeting room
cabinet

会议室
壁橱

Lukov Pereulok Apartment
Lukov Pereulok 公寓

建筑师 A. Kozyr, I. Babak
位置 Moscow, Lukov pereulok
结构工程师 A. Plotnikov, I. Selivestrov
面积 300 m^2
完工日期 2003

类别 apartment
摄影师 V. Kuznetsov

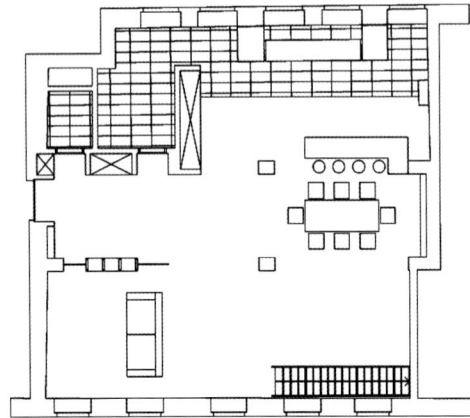

floor plan
second floor, glass staircase
and entrance to the bedroom

平面图案
二楼，玻璃楼梯细部
卧室的入口处

This spacious apartment was formed by combining two flats in the two upper storeys and attic of a normal apartment block. The advantages of the loft thus created include not just the majestic view from the roof terrace, but its sheer size. The different levels are linked by clever management of the copious daylight. The materials used also convey transparency and openness. The treads on the staircase on the first floor are made of thin plates of onyx; a frontage of glass bricks divides off the bedroom. The architect's »handwriting« can be seen in the selection of surfaces: black polished concrete, dark woods, and of course metal – which, when used for built-in components, has been employed both for construction and decoration.

这套宽敞的公寓由一栋标准的公寓楼顶部的两个高层和阁楼合并构成的双层复式结构。这一阁楼的优点不仅仅是可以从屋顶平台观看壮丽的景色，也有它绝对的空间。不同楼层由充足的日光巧妙的调节连在一起。应用的材料也传递出透明性和开放性。一层的楼梯台阶是玛瑙薄板制成；玻璃砖的正面将卧室隔开。表面精华区清晰可见建筑师"笔迹"：黑色的被抛光的混凝土、暗色的木料，当然也有金属（嵌入部分使用）无论是在建筑结构和装饰方面都得到了使用。

dining and living room
living room, detail
kitchen

餐厅和起居室
起居室，细部
厨房

Apartment in Pushkarev Pereulok
Pereulok 公寓

建筑师 N. Lobanova
位置 Moscow, Pushkarev pereulok
面积 220 m²
完工日期 2006

类别 apartment
摄影师 M. Stepanov

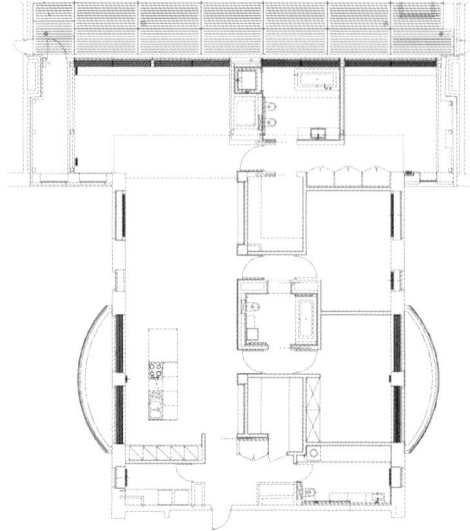

floor plan
living area

平面图
居住区

Every architect should plan a house in white at least once in his working life. That, at least, is what the creator of this apartment says, to whom white gave an opportunity to concentrate on what was essential: space and material. The apartment consists of three completely white rooms. Only the bathroom, with its brown and green tiles, provides a surprising exception. The design concept relies on artificial light, the varied use of which compensates for the lack of any colouring. The lighting units, designed by the architect herself, have less in common with lamps and than with artificial hybrids, alternating between material and dissolution. They give an optical depth to the »white« atmosphere and assert the material nature of the surfaces. Despite its stringency of design, this apartment has nothing of the aseptic purity of minimalism; rather, with its relaxed, light atmosphere, it is reminiscent of Mediterranean rooms.

任何一位建筑师都希望在其设计生涯中设计一所以白色为主题的房子。至少，这一观点是这套公寓的创造者坦言的，对于他来说白色是为集中精力钻研建筑的本质——空间和材料——提供了一次良机。这套公寓包括三间全白色的房间。只有浴室，棕色和绿色的瓷砖为整套公寓平添了令人讶异的例外。设计的理念依靠人造光源完美体现，各种人造光源的使用弥补了所有色彩的缺憾。由设计师亲自设计的照明单体不同于普通的灯，在材料与光线的分解中交错更迭。它们为"白色的"氛围增添了视觉深度，强调了建筑外表材质的真实本性。尽管设计极具严格性，但是这套公寓忽视了最低限的洁净纯度；更确切的说，由于这一点不拘小节，在光形成的氛围中，让人联想到地中海一带的房屋。

Artplay Design Centre

Artplay工作室

ArtPlay Architects   ARTPLAY 建筑师事务所

Artplay Design Centre
Artplay工作室

建筑师 S. Desyatov
结构工程师 A. Bochkov
位置 Moscow, ulitsa Timura Frunze 11/34
面积 10.000 m²
完工日期 2005

类别 office
摄影师 V. Efimov, S. Leontyev, À. Yagubsky

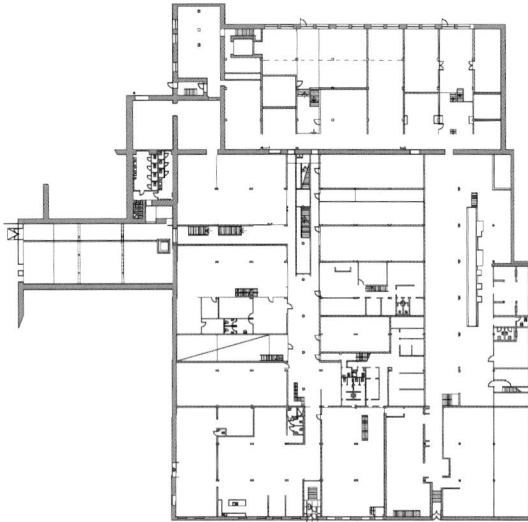

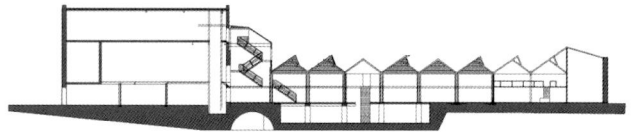

ground floor plan and section
The ArtPlay's central nave is reserved for exhibitions, presentations, and parties.

地下室平面图
ArtPlay的中央广场
被用做展示间和宴会间

This design centre is situated on the site of the former Krasnaya Rosa (Red Rose) silk factory, built in 1875. By the end of the nineteenth century this area, formerly a residential one, was home to various factory complexes, with wooden huts, an office building made of red clinker, a dyeing works, and a large underground store. Some of these historic structures have been revitalised through conversion and redesigned to reflect a new use. The users of this new design centre – 51 in total – all belong to the contemporary architectural and design scene in the widest sense, and it perfectly reflects this discipline's current popularity that all the architectural practices here are located in prestigious offices on the building's first floor. This open »bel etage«, intended for presentation and exhibition purposes, and immediately adjacent to the two cafés, links the general public, in its comings and goings, with what is happening inside the offices and studios.

该设计中心位置设在红玫瑰丝绸厂（建于1875年）的旧址上。直到19世纪末这一地区（从前曾是住宅区）是各种工厂复合建筑的聚集区，有木制的小屋、红色炼砖建成的办公楼、染坊和一座大型的地下商店。这些历史性建筑中的一部分通过改造焕发了新的活力，并且被重新赋予了新的使用价值。

新的设计中心的使用者——共计51人——全部属于当代拥有最宽广的设计理念的建筑设计场所，完美地反映了这个学科当前的流行性，全部的建筑事务所设在这栋楼一层的久负盛名的工作室内。计划作为介绍和展览使用的开放性的"贝尔层"和紧邻的两家咖啡厅在熙来攘往中连接着公众与内部工作室中发生的一切。

91

NEW INTERIOR DESIGN IN RUSSIA

interiors of architecture offices
accomodated at ArtPlay:
Àrt-Blya office detail
passageway flanked by
architecture offices
office interiors, details:
office Arch-4,
office Sergei Tchoban

位于ArtPlay中的建筑办公室内部图

The ArtPlay Design Centre is home for
an architecture office of the same name,
which occupies the space around the
main nave.
office interior, details:
office Project Meganom

ArtPlay设计中心位于与之同名的ArtPlay建筑办公
室腹地，它占据了中央广场的周围空间。
办公室内部，细部
Meganom项目办公室

Yevgeny Ass   YEVGENY 建筑师事务所

Office Building in Chelobityevo
Chelobityevo 办公楼

建筑师 Y. Ass
位置 Chelobityevo
面积 740 m^2
完工日期 2004

类别 office
摄影师 Y. Ass

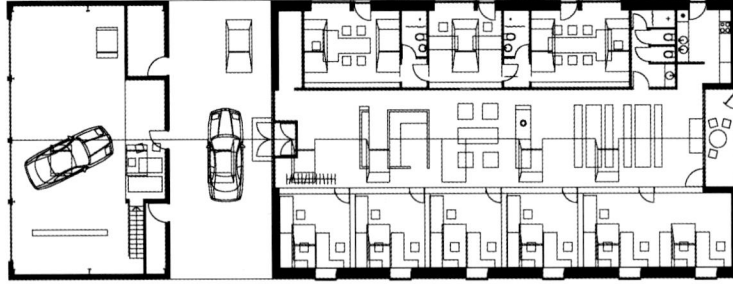

98

floor plan
reception

平面图
接待处

In a dull village marred by disparate development within the commuter belt of the Russian capital, a small firm has proven with its office building that good, functional architecture does not inevitably have to be faceless and joyless. The elongated flat building is reminiscent of a warehouse in its simplicity. It includes garages for company vehicles as well as offices, reception and recreation areas. The employee offices can be found in the southern area, separated by the garages and utility rooms. The central section, which includes the main entrance, was designed to serve as reception and communication area. The modern solid-wood furniture in this section was designed by the architect. The northern section of the building is reserved for the management. The building's external design is the most captivating aspect of this compact, one-storey gabled construction: the brickwork that covers the roof and the entire length of the building on both sides contains a diagonal coloured pattern which lends the simple structure a dynamic feel.

在一座冷冷清清的村庄里（在俄罗斯首都的通勤居住区内全异的发展打破了它的宁静），一家小型的公司已经证明了它的办公楼，那栋优质的、功能齐全的建筑不是不可避免的必须成为一栋毫无个性化的、沉闷无趣的建筑。被延长展开的建筑的简洁性让人联想起仓库。这栋建筑包括停放公司车辆的车库以及办公室、接待处和娱乐区。雇员的办公室设在南部区域，由车库和杂物间隔开。包括主入口的中心区，被设计成为接待处和会客区。在这个区域中现代风格的实木家具由建筑师设计。建筑的北部区域保留作为管理区。这栋建筑的外部设计最吸引人地方是这个坚固的一层高的带有山墙的砖砌建筑（这栋建筑的屋顶和两个侧面全部由砖砌成），它包含了一个斜纹的彩色图案，将简单的结构赋予了动态的美感。

Office Novinsky Boulevard
Penthouse »Scarlet Sails«
RuArts Gallery
Apartment in Ostozhenka District

Büro Novinski 大道
红帆屋
俄罗斯画廊
Ostozhenka区公寓

Atrium ATRIUM 建筑师事务所

Office Novinsky Boulevard
Büro Novinski大道

建筑师 A. Nadtochy, V. Butko, A. Alenicheva
位置 Moscow, Novinski bulvar
面积 400 m²
完工日期 1998

类别 office
摄影师 A. Knyazev

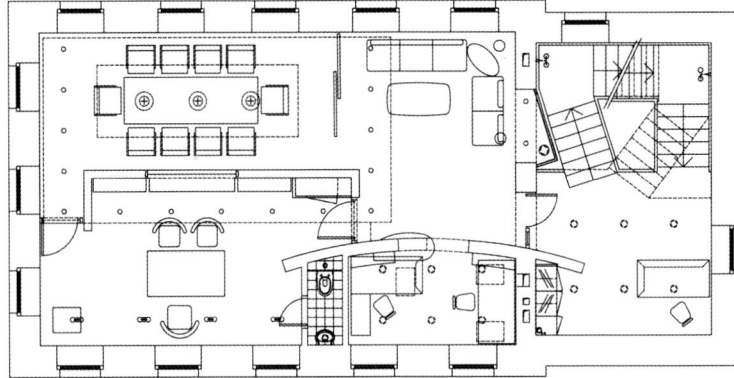

first floor plan
stairway
一楼平面图
楼梯

104

Transparency is the decisive characteristic of this office. All rooms are linked via undisguised glass surfaces – partitions, doors, windows. The interface with the outer world is an existing brick wall, with small, regularly placed windows. This wall has been converted into a projection surface, which changes its appearance and haptic features with the various functions of the individual office areas. It also has a space-generating, architectonic role to play, reconciling atmospheric contradictions. Thus the director's office has been designed as a forbidding steel cage, deliberately reminiscent of a safe in its seclusiveness, forming a crass contradiction with the transparency desired. The interior decoration has been used to dissipate this spatial and material tension, transforming it into a set of positive equivalents.

透明性是这间工作室明显的特征。全部房间由毫无装饰的玻璃平面（如隔墙、门和窗子）连接起来。一堵原有的砖墙形成了与外界的分界面，可以看到一扇扇狭小的、中规中矩的窗子。这堵墙已经变成了一个投影面，每一个独立的办公区根据各自不同的功能将墙的外观及触感做了相应的改变。同时，它也在创造和构筑空间中发挥着作用，使矛盾的氛围变得更加和谐。因此，经理办公室已经被设计作为一个令人畏惧的铁笼，故意让人想起一个保险箱，与渴望中的透明性形成了一个粗糙的矛盾体。这种室内装饰被用来缓解空间与材料间的不协调状态，使其转变成为一系列绝对均等的空间。

106

Penthouse »Scarlet Sails«
红帆屋

建筑师 A. Nadtochy, V. Butko, O. Sokolova
位置 Moscow, Shukino
面积 250 m²
完工日期 2003

类别 apartment
摄影师 A. Knyazev

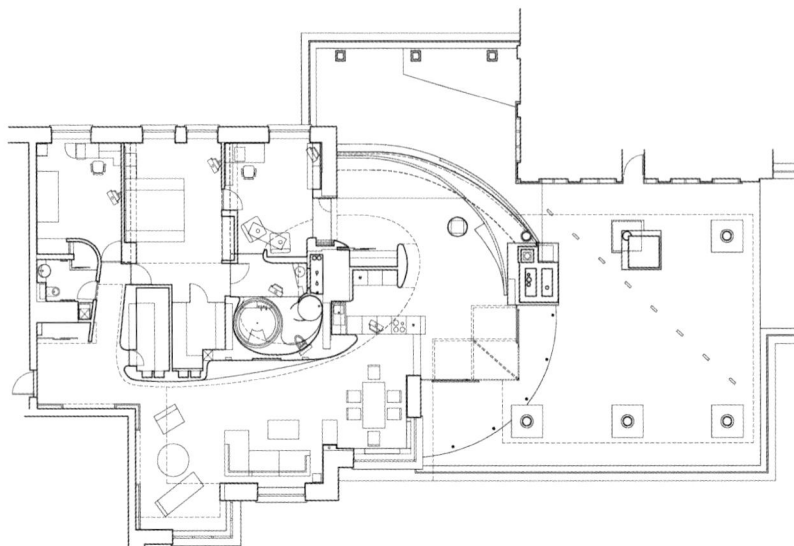

floor plan
penthouse

平面图
红帆屋

The penthouse has access to the roof over which a giant sail is hung. This maritime feature gave the project its name. It consists of an apartment with an original living area of 150 square metres along with a terrace some 80 square metres in size. The renovation has resulted in an increase to the living space. A chimney constructed of poured concrete forms the boundary between the interior and the exterior whereby it belongs equally both to the inside and to the outside. This area is separated by a glass wall which is attached directly to the ceiling and may be opened in the manner of a sliding door. The heart of the apartment consists of the so-called amoeba – an area formed by red textile-clad concrete that accommodates the dressing room, bath, building and media utilities together with a built in kitchen. White is the predominant colour in the living and social areas. Its airy ambiance matches the atmospheric of the inter-penetration of interior and exterior.

帆屋的屋顶上悬挂着一面巨大的帆，该公寓也因之得名。它由150平方米的设计新颖的起居空间和80平方米的露台构成，这种革新增大了居住空间。混凝土浇筑而成的烟囱形成了室内外的分界，它的本身也恰好被内部和外部空间等分。该区域由一扇直达天花板的玻璃幕墙隔开，幕墙可以像滑门一样打开。公寓的中心部分由所谓的"阿米巴"组成——一个由红色织物包覆的混凝土筑造而成的区域，同时也构成了建筑的基本功能区，例如更衣室、浴室、媒体功能间以及组合厨房等。生活区和社交区主体色调是白色。良好的通风与室内室外相互贯通的氛围相得益彰，这也是全部设计的一个特色。总之，这是一个让人能一见倾心的地方。

ATRIUM

109

view from the terrace
view outside
interior with chimney

阳台景观
外景
带有烟囱的室内景观

RuArts Gallery
俄罗斯画廊

建筑师 A. Nadtochy, V. Butko, O. Sokolova, A. Malygin
位置 Moscow, 1st Zachatievski pereulok 10
面积 850 m²
完工日期 2004

类别 cultural facilities
摄影师 A. Naroditski, E. Smirnova, A. Knyazev

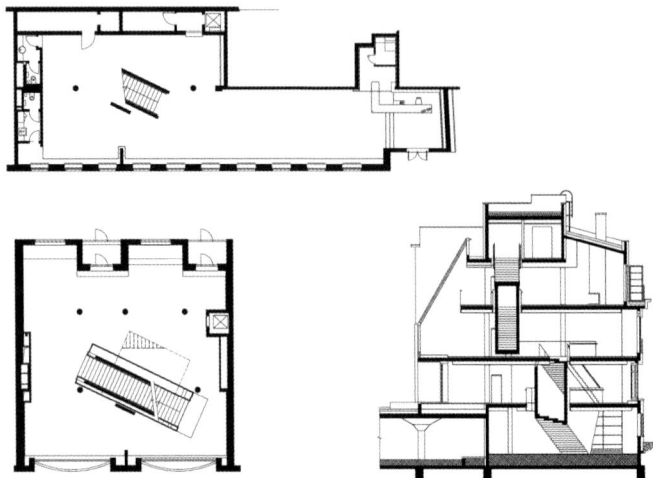

ground floor, and
1st floor plans
section
the gallery's core element:
the staircase

地下室和一楼平面图
剖面图
画廊的核心建筑
楼梯

This gallery in the centre of Moscow intends to be something exceptional. And that's what it is, if its interior design is anything to go by. Instead of a structured circuit on various levels, the visitor finds himself in a vertical structure. In the centre of this interior scheme stands a staircase, around which everything is grouped. It is the gallery's backbone, and if it reminds you of the circuit in Frank Lloyd Wright's Guggenheim Museum in New York, that is no coincidence. But here it does not distract the visitor from the art, but forces him to confront the exhibits. In doing so, the staircase will be at his back. This creates a spatial atmosphere which points beyond the artworks and encourages movement. The interior of this gallery is a social zone, less for solitary contemplation of artefacts – i. e. objects – more for focusing on the visitor, i. e. the subject. It is all about the mutual relation between art and observer, in a spatial organisation which invites interaction.

位于莫斯科中心的这间画廊试图突破传统。如果它的内部设计是任何令人驻足观看的东西，那就是它要表达的。代替多层圆形结构，参观者发现自己处于一个纵向的结构中。设计师在室内中心处立了一个螺旋形的楼梯，每件东西都是环绕它而建。它作为画廊的支柱，是否使您想起了由法兰克·洛伊莱特设计的纽约古根海姆现代艺术博物馆。与其不同的是，在这里没有使参观者从艺术中转移注意力，而是促使他面对展品。当参观者欣赏画作时，楼梯在其背后。这营造了一种重点超出艺术品，并鼓励运动的空间氛围。这间画廊的内部设计成一个社交区，减少了对艺术品的孤独关注——即物体——更多集中于参观者，即主题。在一个引发交流的空间环境中，一切都是为了艺术与观众间的相互交流。

the two concrete cubes of
the staircase
带有混凝土结构的两处楼梯间

Apartment in Ostozhenka District
Ostozhenka区公寓

建筑师 A. Nadtochy, V. Butko, O. Sokolova
位置 Moscow, Ostozhenka district
面积 300 m²
完工日期 2006

类别 apartment
摄影师 A. Nadtochy

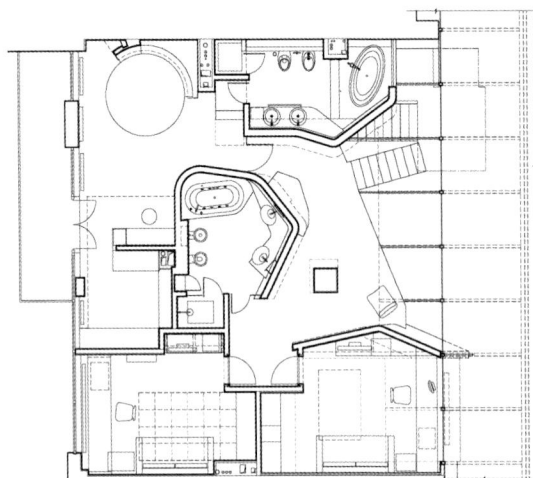

floor plan
staircase

平面图
楼梯间

This 300 square metre apartment extends over two storeys in an older renovated building, and its most attractive feature is its unspoilt view of the newly rebuilt Cathedral of Christ the Saviour with its golden onion domes. This privileged location also served as the starting point when it came to planning the rooms, which open as it were onto the Cathedral. Thus you can enjoy the sight not just from the central living area with its broad windowed frontage, but also from the other rooms in the apartment. This creates an unusually authentic relation between the inner and outer. The interior design makes great use of Corian, steel, comfortable upholstery, glass and wood colours.

这套面积300平方米的公寓位于一栋重新翻修过的老式建筑内，占据了两层的空间。公寓最吸引人的特点是没有被毁掉的，带有金色洋葱型屋顶的，重新修缮的耶稣基督大教堂这一景色。这一特权的位置同时也作为规划内部房间设计时的着手点，尽可能展现的如同站在大教堂之上欣赏景色。因此，您不仅仅可以在中心居住区宽阔的临街窗户前，也可以在公寓的其他房间凭栏远眺欣赏美景。这在内部与外界间创造了一个罕有的、可信的联系纽带。内部设计大量的使用了可丽耐（Corian）、钢材、舒适的室内装潢、玻璃和实木色彩。

117

view from the living room
to the staircase
glass insert in the floor
从起居室到楼梯景观
玻璃嵌入的地板

119

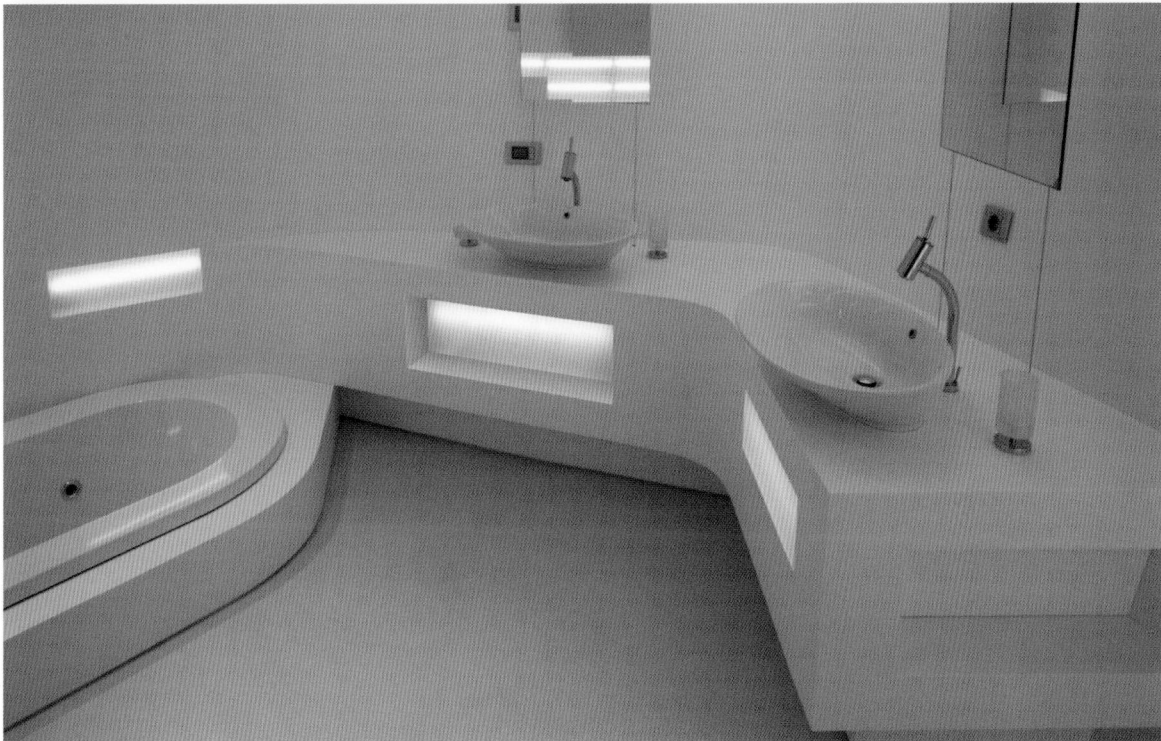

staircase, detail
kitchen
bathroom

楼梯,细部
厨房
盥洗室

Mikhail Vrubel Retrospective
Café »Major Pronin«
Exhibition »Lightness of Being«

杜塞尔多夫艺术画廊
莫斯科大型普罗尼咖啡馆
莫斯科郊外的展览

Yuri Avvakumov   YURI AVVAKUMOV 建筑师事务所

Mikhail Vrubel Retrospective
杜塞尔多夫艺术画廊

建筑师 Y. Avvakumov, A. Kirtsova
位置 Städtische Kunsthalle Düsseldorf
面积 900 m^2
完工日期 1995

类别 cultural facilities
摄影师 Nic Tenwiggenhorn

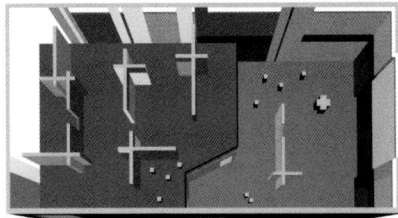

122

3D-visualization
of the exhibition
and the balcony
view from the balcony

展馆和阳台
3D效果图
阳台

The design for the Art Exhibition in Düsseldorf alludes to pre-revolutionary Russia at the beginning of the 20th century and functions on the basis of the art-within-art principle. The spectacular show with art works from Mikhail Vrubel was transported to a fictitious Russian manor house from the turn of the century period, which the architect brought to life within the Art Gallery solely by the use of colour. The exhibition architecture is based on a chronology, which can be traced by means of various projections. The Vrubel paintings were mounted on expansive, coloured walls which themselves form a type of three-dimensional artwork.

杜塞尔多夫艺术展馆的设计指的是20世纪初俄罗斯革命前的设计，并且以艺术中的艺术为基础原则发挥着作用。在世纪之交包括维鲁贝尔的艺术作品在内的展览秀被带到一座虚构的俄罗斯领主的宅邸中，建筑师通过对色彩独到的运用将这个建筑赋予了生命。展馆的展品以纪年为序排列，可以通过不同的投影查询。维鲁贝尔的油画被镶嵌在宽大的、色彩华丽的墙壁上，它们共同形成了一件立体的艺术品。

YURI AVVAKUMOV

124

impressions of the exhibition

展馆印象

Café »Major Pronin«
莫斯科大型普罗尼咖啡馆

建筑师 Y. Avvakumov, A. Bilzho, A. Kirtsova
位置 Moscow
面积 150 m²
完工日期 2002

类别 bar/restaurant
摄影师 A. Kuznetsov

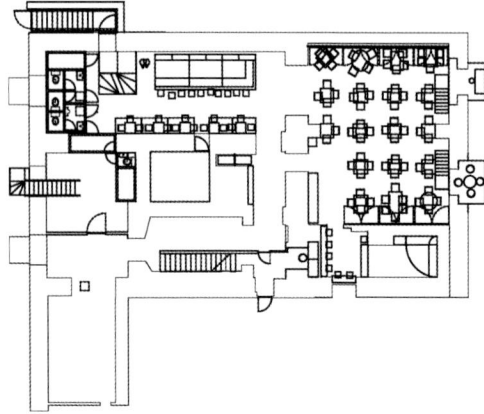

ground floor plan
general view

地下室平面图

The work of this architect contains repeated references to the different stages of Russian-Soviet history. And so the design for this café represents the communist period of Soviet Russia. The structure is based on an imaginary criminal museum from this era. And the gastronomical area too is designed to resemble a prison museum. Open barred doors separate the tables rowed along the wall, from one another, and create niches with a quite unique atmosphere. For larger events, the doors can be folded back making for a larger coherent room space. The exhibited artworks with diverse stylistic leanings lend the interior the requisite museum allusion and at the same time, manifest the deconstructive absurdity of the totality.

在该建筑师的作品中能够多次看到苏俄历史的各个不同阶段的印记。因此这个咖啡馆的设计表现出了苏联共产主义时期的风格。建筑的结构以当时的一个虚构的犯罪博物馆为蓝本。美食区也被设计得如同一间监狱博物馆。敞开的木栅栏门将依墙摆放的咖啡桌彼此分开，创造出一个个氛围独特的小空间。遇到大型的场合，这些门向后折起就会形成一个较大的风格统一的空间。各种风格不一的艺术品为室内平添了必不可少的博物馆印象，同时也证明了解构主义整体的荒谬性。

Exhibition »Lightness of Being«
莫斯科郊外的展览

建筑师 Y. Avvakumov
位置 Pirogovo Resort, Moscow suburb
面积 4.000 m^2
完工日期 2006

类别 cultural facilities
摄影师 Y. Avvakumov

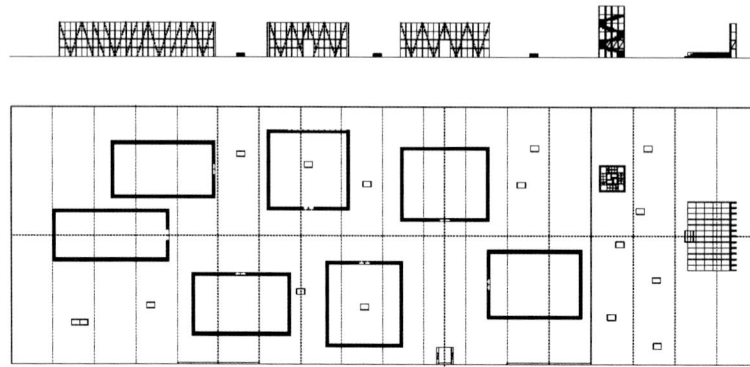

128

section
exhibition floor plan
view into the exhibition

剖面图
展馆平面图
展馆内部图

In the early 1990s two Russian artists, the painters Vinogradov und Dubossarsky came up with the idea of painting pictures and selling them to order in a manner more common with ordinary »goods«. Some ten years later they have become famous due to this concept. The opening of this large-scale exhibition of their works, located in an aircraft hangar near Moscow, was celebrated as one of the social events of the season. The works were housed in seven boxes manufactured from white shelving boards and presented like goods in a wholesaler's. It is hard to envisage a better way of conveying the commercial character of the exhibited works. The cubes themselves form a miniature town: the arrangement of theme pavilions, squares and the viewing tower are reminiscent of the trade fairs showing off the economic achievements of the former Soviet Union, which used to take place in Moscow. The works exhibited here are alluding to socialist realism.

在20世纪90年代初期，两位俄罗斯画家维诺格拉多夫（Vinogradov）和杜伯萨斯基（Dubossarsky）提出给画上色，并且以订单的形式销售的创意，在某种意义上也就是与普通的货物一样公开销售。大约10年后，凭借这种思路他们成名了。他们在莫斯科郊外的飞机库举办了大规模的展览，这次展览也成为该地区本季度著名的社会事件之一。展览由一位知名的设计师担任设计，他将画作放进七个由白色的木板制成的货架样子的盒子中，就如同批发商展示货物一样。很难想象出一个比这更好的方式来传递展出作品的商品特性。一个个立方体自身就形成了一个微型小镇：主题馆、主题广场和瞭望塔的排列不禁让人回想起那些过去在苏联莫斯科地区硕果累累、盛况空前的商品交易会。当有人认为这里展出的作品反映了社会主义写实主义，那么说明这些具有斯大林主义特色的建筑对于表现这种主题是十分恰当的。

YURI AVVAKUMOV

130

Mikhail Belov MIKHAIL BELOV 建筑师事务所

Grand Collection Gallery (Faberge) Boutique
伟大的收藏品画廊（法贝热）精品店

建筑师 M. Belov
位置 Moscow
面积 55 m²
完工日期 2004

类别 retail
摄影师 M. Belov

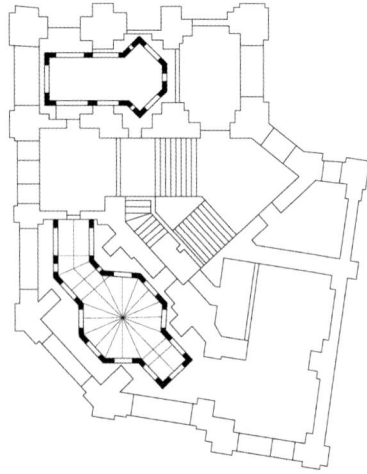

134

ground floor plan
»red room«, detail
following pages:
showroom

地下室平面图
"红色房间"，细部
下一页：展示间

The address is itself already a challenge: the Red Square in Moscow, directly opposite St. Basil's Cathedral. This is where the rooms of an historic, monument-protected house were remodelled into an exclusive antique gallery. Major interventions in the building substance were not possible for reasons of conservation. This is why the architect decided to make the most of his imagination in the realm of interior design. The »white rooms« and »red rooms,« with their own respective atmosphere, form an exciting enfilade. The areas in white breathe an almost clinical minimalism, while the adjoining »red rooms« luxuriate in Tsarist pomp with gold and purple, antique columns and fabric draperies. The historicising details such as column capitals and acanthus décor conceal, at times, the most modern lighting and security technology.

店铺本身所处地点就已经形成了一个挑战：莫斯科红场，正对面就是圣巴索大教堂。这里就是这栋具有历史意义的、需要被保护的遗址性建筑的房间被改建成为一处唯一的古董画廊的地方。建筑材质的主要介入是不可能以保护为理由的。这就是建筑师决定充分利用其在室内设计领域的高超想象力的原因。"白色房间"和"红色房间"以及各自的氛围形成了一个令人兴奋的纵向落差。白色的区域散发出几乎如同病房中才能感觉到那种极少主义的气息，而毗邻的"红色房间"大量使用的金色和紫色，古董柱子和织物装饰品奢华程度几尽俄国帝制全盛时期的景象。那些再现历史的细节，如柱子的顶端部分和叶形装饰板，不时的遮隐住最具现代化气息的照明装置和安全技术。

MIKHAIL BELOV

135

136

President's Cabinet. Installation
Wine Shop
Grand Cru Wine Shop

总统的工作间
02级优质葡萄酒店
莫斯科高级葡萄酒馆

Bernaskoni BERNASKONI 建筑师事务所

President's Cabinet. Installation
总统的工作间

建筑师 M. Belov
位置 Moscow
面积 55 m²
完工日期 2004

类别 retail
摄影师 M. Belov

Interior view of the cabinet:
Information is the main
spatial construction element

壁橱内部
信息是空间
结构的核心元素

This design to a certain extent is architectural science fiction. This attempt to anticipate a workroom from the year 2020 has resulted in a study of the transformation of room functions under the influence of continually developing digitalisation. Information is everything; everything is information. The central element of this workroom is consequently the giant plasma screen with information flitting over it and changing at rapid speed, creating constantly updated light and colour ambiances. The lighting has also been designed according to this principle. Instead of standing or work lights, there are only light panels that are both lamps and elements, which form the spatial character. Only the fittings are reminiscent of the traditional working and living world. They have the task of fulfilling the never-changing demands that people have of a room: a chair in which to sit, a desk at which to work. These appear to be elements that will endure.

一定程度上这种设计堪称建筑史上的科幻小说。在飞速发展的数字化浪潮的影响下，尝试设计一间2020年的工作室已经引发了一项关于房屋功能变革的研究。信息就是一切；一切就是信息。因此工作室的核心元素是能够迅速显示和刷新信息的巨大等离子屏幕，它营造出不断幻化的色彩和光的氛围。灯光也是根据这一原理设计的：代替普通的或者工作照明的只有光板，既是用于照明的灯具也是形成空间特征的元素。只有家具还能给我们带来对传统工作和生活方式的回忆。它们拥有着满足人类需求的使命，一项人类需要房子带给我们的亘古不变的需求：用来休息的椅子，用来工作的桌子。这些体现着一种永恒。

Wine Shop
02级优质葡萄酒店

建筑师 B. Bernaskoni, E. Lyubavskaya, D. Mikheikin,
O. Treivas, A. Bystritski
位置 Moscow, Leningradski prospekt 50
面积 110 m^2
完工日期 2007

类别 retail
摄影师 Y. Palmin

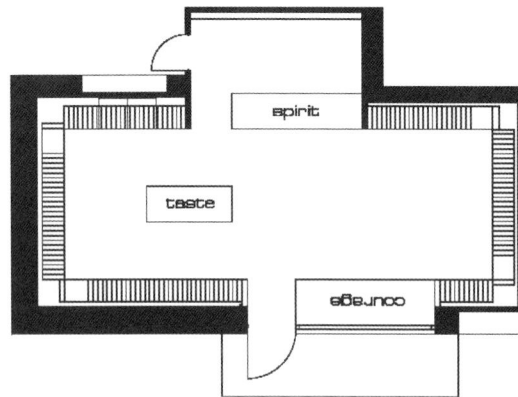

floor plan
interior view

平面图
内部空间

It is one of those small, high-quality shops which manage to buck the prevailing trend, relying on quality alone. This applies not just to the goods on sale, but to the shop's design. Instead of shop windows, here there are wine racks; the counter doubles as a bar. The walls of the small room have been covered with a structure of three-dimensional cells, making it possible to keep the wines on offer in a strict order. The shop's exclusive character can be seen here, if not before, for there is room for only one bottle per cell. The wines are arranged in a sort of archive system, by origin, type, year and price. The table between the displays would be well suited to a library. But here it serves to accommodate no more than glasses. Nor is it by any means as quiet here as it would be in a study. When the proprietor turns on the bright light and puts on some music, the interior colours, which at first seemed low-key, reveal themselves as bright, indicative elements – and then this fine wine shop turns into a blatant amusement outlet.

它是一间与众多小型的高品质葡萄酒店一样凭借酒的品质努力跻身于流行行列。这不仅仅涉及到销售的货物，而且也包括店铺的设计。代替店铺的窗户，那里放置了葡萄酒架；收银台也成了一个酒台。小房间的墙壁被一个众多立体的小空间结构覆盖，使得葡萄酒能够保持一个严格的次序不断的被供应。这里可以见到这间店的独特的特色（如果在此前没有注意到），每一个小格子只有容纳一瓶酒的空间。葡萄酒依据它的产地、类型、年代和价格归档存放。展示架之间的桌子非常适合于图书馆使用。但是在这里它只是作为玻璃制品。无论如何它也不会像书房一样成为安静场所。当经营者打开明亮的灯，播放一些音乐，刚开始似乎是低调的内部色彩，展现它们自己如同耀眼的直陈元素。欢乐的气氛达到了顶点——这家优质葡萄酒店变成了一处喧嚣的娱乐消遣处。

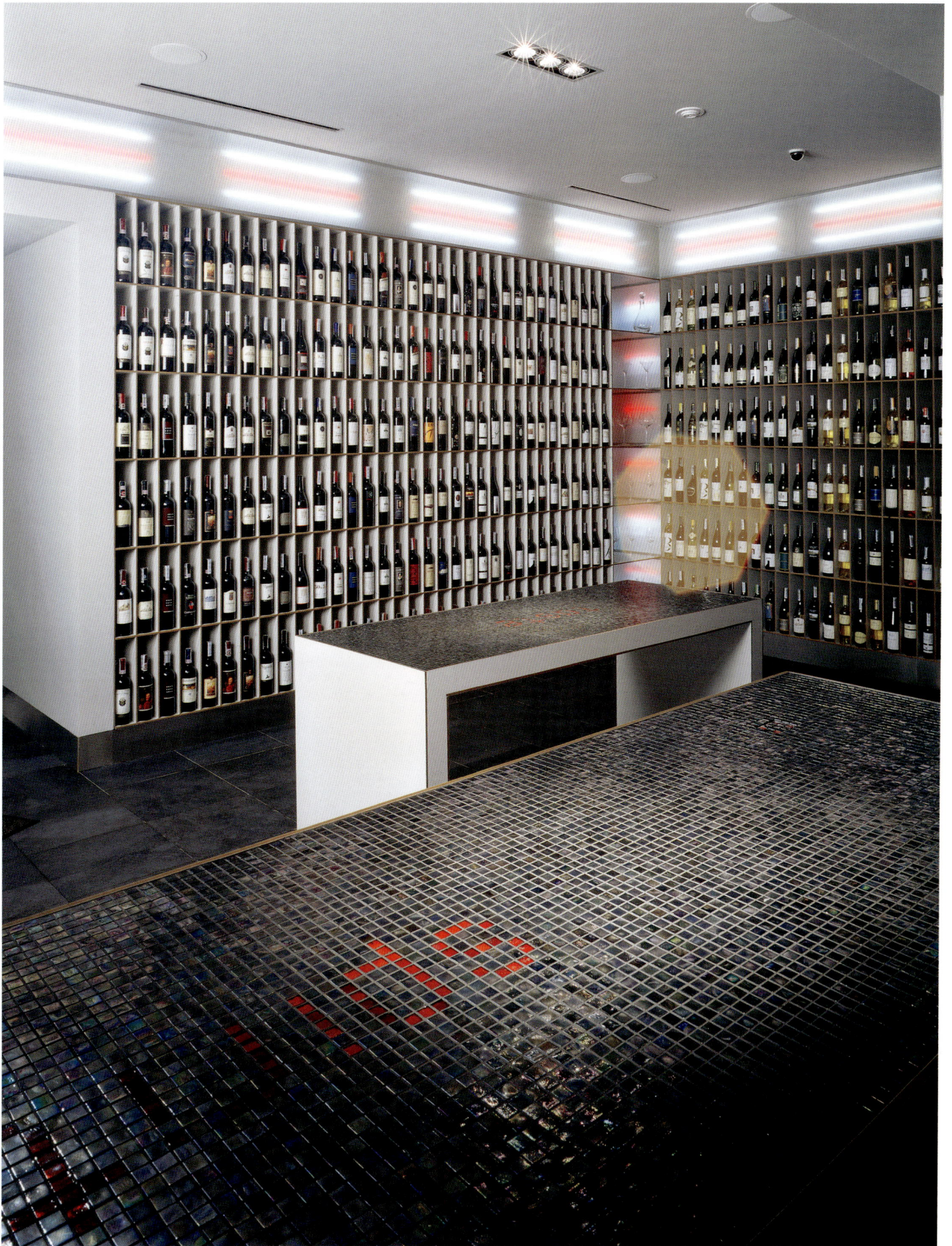

144

Grand Cru Wine Shop
莫斯科高级葡萄酒馆

建筑师 B. Bernaskoni, E. Lyubavskaya, D. Mikheikin,
O. Treivas, A. Bystritski
位置 Moscow, Leningradski prospekt 50
面积 103 m^2
完工日期 2007

类别 retail
摄影师 Y. Palmin

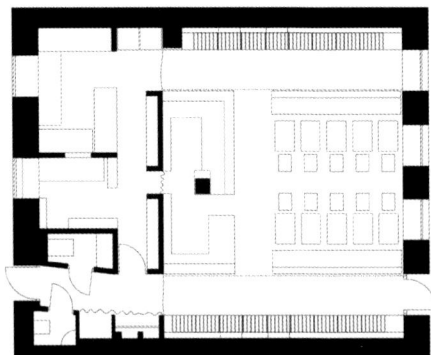

floor plan
view of the wine shelves
and the glass curtain

平面图
酒架一览
玻璃制窗帘

The wine store with the programmatic name »Grand Cru« is located in ulitsa Malaya Bronnaya in Moscow and is both a retail store and a bar. This coupling of functions is realised by means of the Matryoshka Principle: the bar is, so to say, inserted into the store. During daytime hours, the two areas are separated from one another only by a flexible glass dividing-wall. Introversion and intimacy distinguish the bar from the open store with its attractions aimed at passers-by. The lowered stucco adorned ceilings import a restrained, private atmosphere. Guest can take a seat on the metre-wide sofas in front of which low tables are arranged. The glass dividing-wall can be transformed into a sparkling backdrop by the use of special lighting; heavy, riveted leather chairs are rowed along the counter. The technically cool visuals of the modern shelf construction, and the decidedly modern language of form, ensure that the atmosphere is never plushily complacent however.

设计这间葡萄酒馆时，建筑师遵循了一个独特的主题。每个葡萄酒馆因其各自规模的大小抑或内部装潢布置的不同而显得千差万别，同时它们又无一例外地具有共同的特有标志，如照明布置、冷色调材料和独特酒架设计。名字中带有"高级"字样的这家葡萄酒馆坐落在莫斯科的Malaya Bronnay，它既是一家葡萄酒零售店，同时也是一家酒吧。这种双重功能的设计灵感来自于俄罗斯套娃，酒吧可以说是插在酒馆之中的。白天，这两块区域由一个活动的玻璃隔墙彼此分开。区别于门庭大开的店铺，酒吧的内向性和私密性吸引着过路人的目光。低矮的灰泥装饰的天花板营造了一个隐秘的氛围。客人可以坐在宽大的沙发上，一些低矮的桌子摆放在沙发前。通过特殊灯光的照射，可活动的玻璃隔墙形成一面闪闪发光的背景。依着柜台摆放着厚重的、铆钉紧铆的皮座椅。现代的搁架式结构给人带来清爽的感觉，再加上外在形式上干净利落的时尚表达方式，置身其中不会感到丝毫的奢华溢满。

147

148

wine shop
wine racks

酒馆
酒架

glass curtain
wine bar

玻璃垂帘
酒吧间

Guelman Apartment
Restaurant »Ulitsa OGI«
Apshu Café – Club
Pavilion for Vodka Ceremonies

Guelman公寓
Uliza OGI 餐馆
Apshu俱乐部
伏特加典礼大厅

Bureau Alexander Brodsky
BUREAU ALEXANDER BRODSKY 建筑师事务所

Guelman Apartment
Guelman公寓

建筑师 A. Brodsky, J. Kovalchuk
位置 Moscow
面积 150 m²
完工日期 2000

类别 apartment
摄影师 Y. Palmin

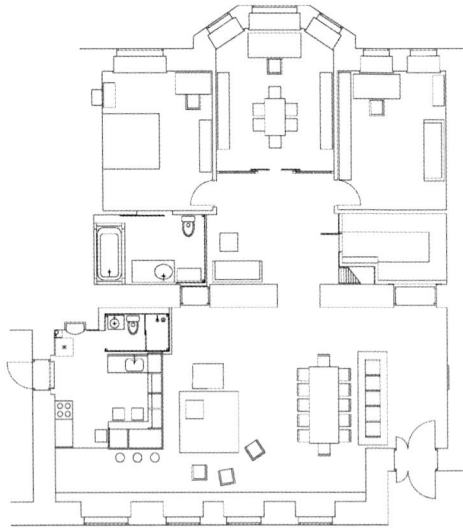

152

floor plan
living room

平面图
起居室

This apartment is situated in a historical apartment block dating from the nineteenth century. The interior has been divided by a bearing wall, which has three openings. While the entrance hall, living room, kitchen and guest bathroom are located in the front section of the apartment, the rear part contains two bedrooms, a study, the bathroom and a wardrobe. The architect's idea of making the interior space seem as open as possible was all the easier to put into practice given the apartment's structural characteristics. The apartment has two spacious windowed frontages, which provide splendid vistas, thus contributing to the open atmosphere desired. The bathrooms and the 2.10 metre high wardrobes were specially constructed for the apartment and are made of glass and steel. The cool nature of these materials is emphasised by the lighting. The bathrooms are lit by lamps fastened above a glass ceiling.

这套公寓位于一栋19世纪的历史上著名的公寓楼内。它的内部由一堵承重墙分开，拥有三个出口。前厅、起居室、厨房和供客人使用的浴室位于公寓的前部。公寓的后部包含两间卧室、一间书房、浴室和一个衣柜。建筑师室内空间的创意（似乎尽可能的开放）是将公寓结构特点赋予的方便性付诸实践。这套公寓有两个宽敞的带有窗子的临街面，在那里可以看到壮丽的街景，因此形成了期望中的开放氛围。浴室和1.10米高的衣柜是为这间公寓特殊定制的，由玻璃和钢制成。灯光加重了这些材料的冷漠淡然的本质。浴室因为玻璃天花板的灯而变得更加明亮。

154

155

Restaurant »Ulitsa OGI«
Uliza OGI 餐馆

建筑师 A. Brodsky, J. Kovalchuk
位置 Moscow, ulitsa Petrovka
面积 350 m^2
完工日期 2002

类别 bar/restaurant
摄影师 Y. Palmin, A. Naroditski

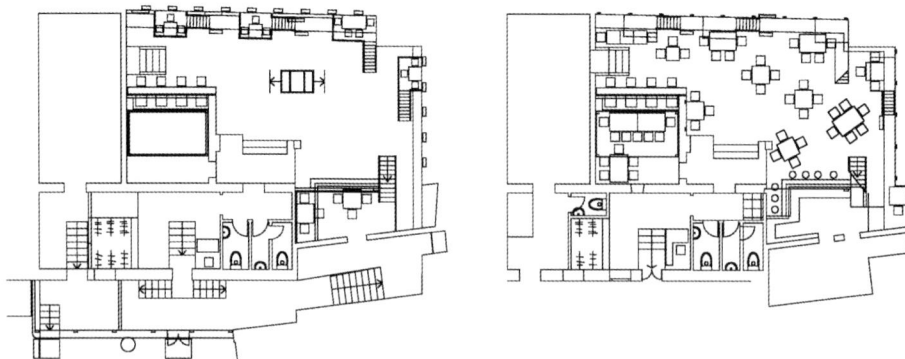

156

first floor plan,
entresol floor plan
dining hall

一楼平面图
夹层平面图
餐厅

The OGI Restaurant was designed after its architect had given up design work for fifteen years, during which time he worked mainly as an artist. The interior, with its emphasis on decadence, reflects strongly Romantic tendencies. As a starting point for his design, a ruin was more or less literally what he found: the dilapidated rooms needed complete reconstruction. After all earlier fixtures had been removed, nothing remained but naked brick walls, plus floors and ceilings of concrete; materials, in other words, which actually belong outside, in the urban landscape. It was by this aesthetic that the architect allowed himself to be guided, designing the restaurant areas like public streets. Interior walls have become facades, whose fire escapes lead to mezzanine floors, used to increase the restaurant's capacity. The central dining area reminds one rather of streets in Venice, or an inner courtyard in Georgia, while the second floor has been kept quite simple and neutral, with coloured walls and wooden floors.

OGI餐馆是在设计师放弃设计工作15年后的作品，此次他主要作为一名艺术家参与其中。重点表现颓废思想的室内设计风格强烈地反映了浪漫主义趋势。作为设计的出发点，他找到了一点遗迹：需要全部重建的坍塌房屋。在去除早期固定的附着物后，除了裸露的砖墙、地板和混凝土天花板外没有其他东西留下。换句话说，实际上这些材料在市区内的景观设计中用于外部的修建。根据这个美学标准的指导，建筑师设计的餐馆区如公共街道一样。内部墙壁已经成为了正面，防火梯通向通常用于增加餐馆内部容量的中层楼。中心就餐区使人会想到威尼斯的街道或者是乔治亚州的一间内部庭院，而二楼彩色的墙壁、木质的地板保留了相当简朴和中性的设计。

interior, details
室内景观，细部

bricks as wall ornaments

砖制墙饰

159

Apshu Café – Club
Apshu俱乐部

建筑师 A. Brodsky, J. Kovalchuk
位置 Moscow, Klementovski pereulok
面积 300 m^2
完工日期 2003

类别 bar/restaurant
摄影师 Y. Palmin

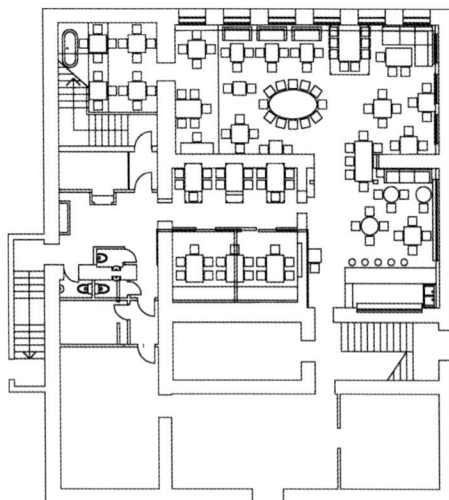

floor plan
view from the entrance

平面图
入口

The basement of a nineteenth-century apartment block has been turned into a club. The attractive architecture of the premises – a series of rooms, all of different sizes, grouped to form a half circle – has been used as the basis for a very idiosyncratic interior design, which might be called a sort of voluntary raw minimalism. The brick walls are unplastered, cables and pipes along the ceiling are uncovered. Big old windows have been painted white, to prevent visitors having the feeling of being underground. There are a number of curious things which strengthen the impression of a fantasy world even further. Thus old curtains conceal an intermediate level, and the wall panelling in the back room is of simple plywood – you normally find elements such as these in a datcha. In the bar bits of old junk are scattered around between the counter and the tables: bookshelves, a large bed, an enamelled bath.

19世纪公寓楼的地下室已经被改造成为一间俱乐部。这栋建筑吸引人的结构是一系列的房间，大小不一，成半圆形排列——这些房间已经被作为基础部分创造一种完全异质的室内设计，这种设计可以被称为随意的最简单派艺术设计。砖墙没有抹灰，沿着天花板的电线和管道裸露着。又大又旧的窗户被涂成了白色，这是为了防止客人有身在地下的感觉。还有许多奇怪的东西，更进一步增强了科幻世界的印象。旧窗帘隐藏着一个中间层，并且里屋墙壁镶板由简易的夹板制成——通常您会在郊外的别墅发现这种材料。在酒吧内少量的旧物散放在收银台和桌子之间：旧书架、大床、一个瓷釉浴室。

entrance area
Interior, details

入口
细部

163

Pavilion for Vodka Ceremonies
伏特加典礼大厅

建筑师 A. Brodsky
位置 Klyazma near Moscow
面积 12,5 m²
完工日期 2003

类别 bar/restaurant
摄影师 Y. Palmin

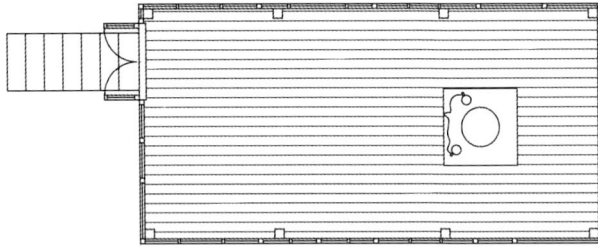

floor plan
view from outside
following pages:
interior

平面图
外部展示图
下一页：
内部结构

Alexander Brodsky long dreamed of building a pavilion for vodka ceremonies. Two factors enabled him to make this dream come true. One was the demolition of the Butikovsky factory in Ostozhenka and the other was the annual Klyazma festival of arts. The festival organisers placed a piece of land at the edge of the forest at Brodsky's disposal and the old factory supplied him with unique material in the form of pre-1917 window frames. In the architect's hands, the frames became a universal building element, forming both load-bearing structure and cladding. The extensive glazed areas do not detract at all from the intimacy of the vodka ceremony because an old coat of white paint renders the windows fully opaque. During the daytime, light filters through the matt windows. When evening falls, candles are lit in the small building and a warm glow starts to emanate. This attracts all devotees of the ceremony, which is very simple and yet ritualistic. Two people stand, drink and feel in complete harmony with themselves and each other.

Alexander Brodsky长久以来的梦想就是为伏特加典礼设计大厅。由于Butikovsky工厂的拆除和一年一度的Klyazma节的缘故，使他终于有机会实现梦想。节日活动主办方在森林的边上设置了一块空地，旧的建筑材料为他提供了原料的支持。在建筑方面，旧的窗框成了建筑结构的主要元素。在白天光线透过窗户进入室内，在晚上，烛光散发出温暖的气息。这一设计受到活动爱好者的关注，人们可以自在的在这融洽的气氛中品尝美酒。

National Jewish Theatre
Prechistinka Apartment
Apartment »Rotonda«

犹太人剧院
威尼斯公寓
Rotonda公寓

Mikhail Filippov   MIKHAIL FILIPPOV 建筑师事务所

National Jewish Theatre
犹太人剧院

建筑师 M. Filippov, T. Filippova, C. Belyavskaya,
T. Deniskina, V. Puk, M. Sharapova
位置 Moscow
面积 500 m²
完工日期 1996

类别 cultural facilities
摄影师 A. Knyazev

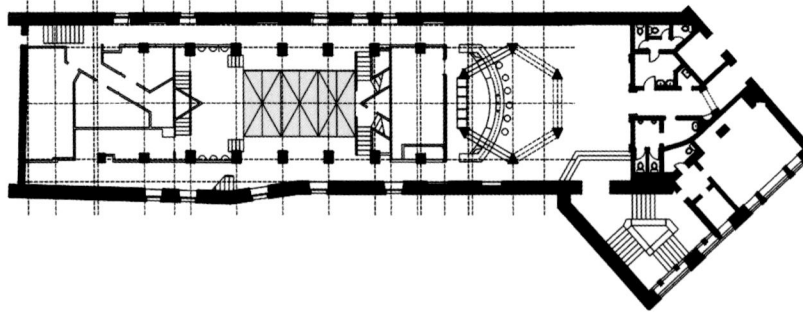

170

floor plan
interior, detail
平面图
室内景观，细部

If one were to attempt to translate the variety of Jewish culture into architectural forms, the temptation would be great to make use of religious symbols and a rich repertoire of ethno-romantic elements – and to reproduce clichés in this way. The architects of the Jewish Theatre in Moscow were surely aware of this risk, for they decided upon a kind of architecture that is clearly inspired by Palladio's »Olympico« Theatre and is above any suspicion of mere folklore. The classical, strictly symmetrical space with the audience terraces, stacked towards the stage like a gallery, is arched by an impressive ceiling construction. In the outlines and forms of the installations, the schooled eye is able to discover abstract interpretations of the old symbols of Jewry: the Star of David and the menorah.

如果有人想要尝试在建筑形态中阐释犹太文化的多样性，那么这种诱惑将会大量的利用宗教符号和包含丰富的浪漫的民族元素的剧目——并且以这种方式大量的复制。莫斯科犹太剧院的建筑师们确实知道这一个冒险，因为他们决定设计一座明显受到帕拉蒂奥"奥林匹克"剧院启发的建筑，并且要超越纯民间传说的任何怀疑。古典的、严格对称的空间以及观众的看台，如同一条长廊一样垒向舞台，宏伟的天棚构成拱形的穹顶。在装置物的轮廓和形态中，受过训练的观察者是能够观察到古老的犹太符号在其中的抽象阐释：大卫之星和大烛台。

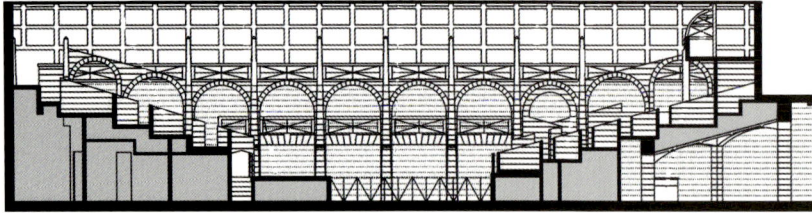
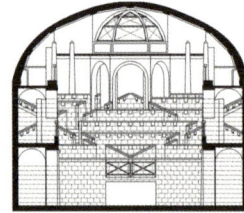

172

MIKHAIL FILIPPOV

Prechistinka Apartment
威尼斯公寓

建筑师 M. Filippov, E. Stroganova, M. Dogadin, V. Cheskidov
位置 Moscow, ulitsa Prechistinka
面积 292 m²
完工日期 2003

类别 apartment
摄影师 A. Naroditski

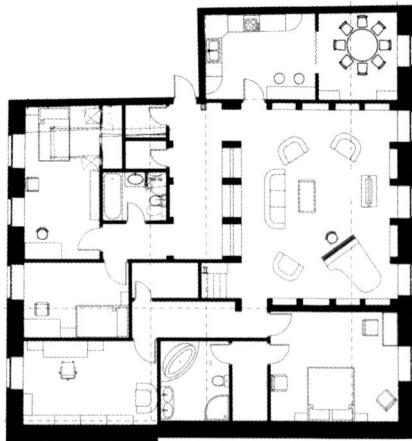

174

floor plan
living room

平面图
起居室

This apartment, in the very heart of the Russian capital, covers the whole area of a former attic. The rooms have been organised around a main hall of two storeys, which is literally flooded with light. The high, arched ceiling, the walls, and the wooden flooring provided the pattern for the whole interior, a pattern evolved with a mathematical consistency, irrespective of the element involved – be it corridors, bookshelves or windows. The proportions are classical. Wide spaces suggest associations with the splendours of Roman architecture, such as basilicas and baths. The romantic-theatrical potential of the interior is plain to see. Large notches in the arched wall are part of the deliberate effect, similar to the trompe l'oeil mirror – a detail which can be found in many of this architect's projects. Bedrooms and study are on the second level and are linked by a gallery, from which, through the slits in the wall, you can look down on the great hall.

这套公寓位于莫斯科中心的一栋18世纪的大型住宅楼内。这一具有城市化的地理位置在起居室的设计中也有所反映。它很容易使人联想到城市迷宫，也是故意如此。从酷似大厅的中心入口处可以到达所有的房间；宏伟的拱门标志可以通向公寓的内部。这些通道引人注意的是它们独特的位置：它们指向的方向并不是朝向它们各自通向的房间，而是朝向公寓外墙。这样您能直接感受到平面图的不规则性。如公寓名字恰恰表明的一样：在建筑师的设计中，他依照威尼斯令人气愤的街道线路为指导，那种很少有正确的角度指向线路，并且总是不规律地通向小巷深处的灌木丛中。其他市内的细节被应用在室内设计中。巨大的门厅是一个古代街道的室内版，其周围是两层高的房子。被精心设计出的错位画的效果使它几乎不再可能告诉这个通道源自何处去向何方。

MIKHAIL FILIPPOV

section
balcony

剖面图
阳台

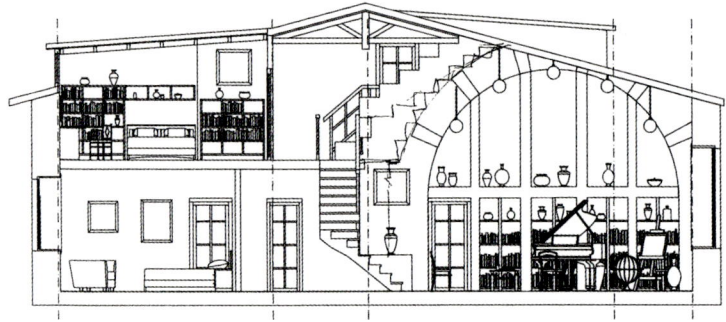

176

MIKHAIL FILIPPOV

Apartment »Rotonda«
Rotonda公寓

建筑师 T. Filippova, E. Alyokhina, T. Deniskina, V. Cheskidov
A. Tarasova, L. Ryazantseva, A. Filippov
位置 Moscow
面积 330 m²
完工日期 2005

类别 apartment
摄影师 E. Luchina

178

main hall
大厅

The design of this apartment needed to reflect a situation which was both advantageous and complex. The interior is bounded on the one side by a five-metre high wall, and on the other by a completely glassed-in frontage. The windowed facade gives the apartment a view over a beautiful park, and this popular location served as a source of inspiration for the design. For an architect there is always a tension between every interior and its urban context – a tension based nevertheless on a fruitful contrast, which can be exploited when planning interiors. In this case, thanks to the proximity of a green urban area, the result was a park pavilion. The architect held strictly to the formal language of classicism. All built-in elements are harmoniously proportioned in a vertical rhythm and given classical details. Only the modern lighting scheme and contemporary furniture make a break with the traditional order. In this classical context they appear as transient elements, dependent on fashion and the passage of time.

这套公寓的设计需要表现出一种既优越又复杂的环境。内部结构受限于一边一堵5米高的墙和另一边完全由玻璃围住的临街面。有窗的正面可以俯瞰莫斯科美丽的公园，这个受人们喜爱的位置为设计提供了灵感之源。对于建筑师来说，每一栋房子的内部结构与周围的环境之间总是存在一个紧张区——然而以成功的对比为基垫，紧张区也可以在内部结构规划时被开发和利用。由于临近城市绿地，在这种情况下，最终决定建造一座园中楼阁。建筑师严格地遵守古典主义的建筑形式语言。全部内置的元素都是以垂直的节奏和谐地对称，再现了细节处古典的韵味。只有现代的照明装置和当代的家具打破了传统的秩序。在这样的古典环境中依赖于时尚和时间隧道，它们作为瞬间的元素呈现。

MIKHAIL FILIPPOV

180

Restaurant »Chetverg«
Café »Belye Stolby«
Restaurant »Khmel i Solod« and Cocktail Bar »Shot«

"星期四"餐馆
Belye Stolby餐馆
Hleb ＋ Solod餐馆

Andrey Gurari   ANDREY GURARI 建筑师事务所

Restaurant »Chetverg«
"星期四"餐馆

建筑师 A. Gurari
位置 Moscow, Ryazanski prospekt 34/2
面积 420 m^2
完工日期 2004

类别 bar/restaurant
摄影师 K. Ovchinnikov

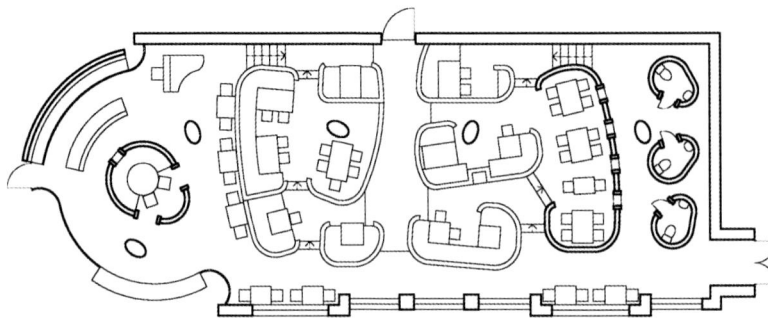

ground floor plan
interior
following pages:
general view

地下室平面图
内部结构图
下一页：
全景图

This new restaurant is situated in a two-storey building, directly fronting Ryazanski prospekt, where there is a large amount of traffic. Through the windows the guests have a view of the motor-way – an odd thing to enjoy, per-haps, though not without its charm. Particularly when you are in a room which itself has an atmosphere beyond the merely everyday and, in its deliberately quiet, monochrome colouring, contrasts strongly with the stri-dent goings-on outside. Here everything is grey. But clever light-ing and use of different materials produce an unexpected variety, showing that grey is a colour of many nuances. The chairs are spread like islands over the room, which is divided with only four par-titions. The planners achieved something unusual with their con-crete surfaces. Directly after the concrete was poured, and before it dried, sheets of plastic were placed in it, giving the walls a deli-cate, almost skin-like touch.

这间新的餐馆位于一栋两层的建筑内，直接对着交通繁华的Ryasanskoer大街。透过窗户客人可以看到高速公路上的景色——大概是一件可供娱乐的怪东西，虽然不是没有吸引力。特别是当你处在一间气氛与往日不同的房中，在它蓄意营造的安静、单色调的氛围，而外界刺耳的嘈杂与之形成了鲜明的对比。这里的一切都是灰色调的，但是灵动的光线和多种材料的运用共同创造出一种出人意料的变化，让人感到灰色也是一种充满了微妙之处的色彩。
房间被分成四个区，其内散布的椅子如同一个个小岛一样。设计师让混凝土的表面有不同寻常的感觉，在混凝土浇筑之后硬化之前，在其表面敷上塑料板，这让墙体表面产生一种如皮肤般细腻的感觉。

ANDREY GURARI

185

186

Café »Belye Stolby«
Belye Stolby餐馆

建筑师 A. Gurari
位置 Moscow, ulitsa Biryuzova 19
面积 140 m^2
完工日期 2005

类别 bar/restaurant
摄影师 K. Ovchinnikov

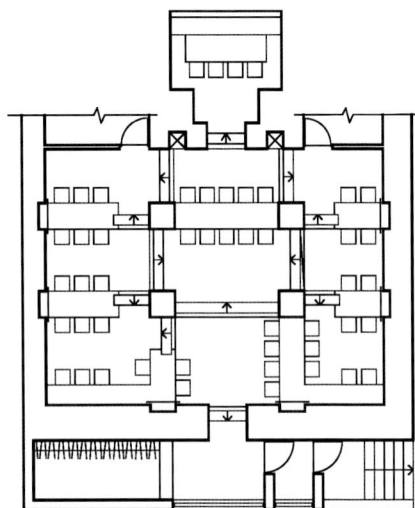

plan
dining room
following pages:
banquet room

平面图
餐厅
下一页：
宴会厅

It really did look as if architecture had finished with white modernism in all its varieties. Yet this restaurant makes every effort to exploit the possibilities offered by this non-colour. Whether a primal condition or the height of perfection – this was not a question which interested the architect here one bit. The conflict between experimental design and ascetic modernism is an area he has not even attempted to enter, and he treats white as pure luxury. In this way he has created a space which quite deliberately relinquishes all simplicity and appears just as complex as it really is. Different areas in the restaurant are characterised by different materials and geometrical structures. The whitewashed brick walls are far from perfect – their uneven texture gives an almost homely feel. All the furniture, and the restaurant accessories, were designed by the architect himself, thus creating a coherent atmosphere, harmonious in every context.

这间咖啡店种类繁多的设计无不体现着白色系的现代主义风格。是的，这家餐馆竭尽所能探索了无色设计风格的可能。是否由于最初的条件或是要求的过于完美——这并不是一个使建筑师感兴趣的问题。试验性设计和苦行现代主义之间的冲突是一个设计师甚至并不尝试踏足的范围，他将白色视为纯粹的奢华。这样他创造了一个空间，一个故意摒弃所有的简洁朴素，表现出的恰恰是空间真正的复杂性。餐馆不同区域使用的不同材料和几何结构使其产生不同的特色。涂有白色涂料的砖墙远非完美，但是凹凸不平的质地就像家一样。所有家具以及餐馆的装饰物，由建筑师本人亲自设计，因此营造了统一的氛围，无处不和谐。

ANDREY GURARI

Restaurant »Khmel i Solod« and Cocktail Bar »Shot«
Hleb ＋ Solod餐馆

建筑师 A. Gurari, E. Gurari
位置 Moscow, ulitsa Aviamotornaya, 10
面积 1.500 m²
完工日期 2006

类别 bar/restaurant
摄影师 V. Efimov

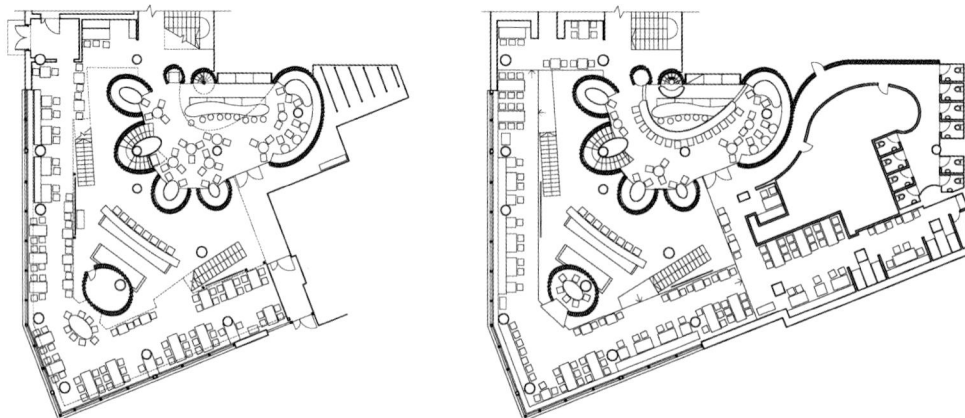

192

first floor plan,
balcony plan
view of the cocktail bar

一楼平面图，
阳台平面图
鸡尾酒吧

Beer and cocktails, as we all know, are not mutually exclusive. But these two drinks are different, at least, when it comes to the atmosphere which they suggest. So it was quite a daunting task to combine a beer and cocktail bar in one room. The architects decided to use the nut-shell principle. Six-metre high hollow cylinders of brick divide the two bar areas. The outer edge of these massive cylinders provides the background to the beer bar, while the inner casing arches protectively around the cocktail area. This produces two distinct zones, each with its own atmosphere: an open, sociable beer bar, and the intimate, private atmosphere of a cocktail bar. The beer bar, adjacent to the front of the building, covers two storeys. Its classic furnishings – simple coffee-shop chairs and dark oak tables – correspond to the aesthetics here of ceiling-high steel gratings and rusty pillars. The cocktail bar, on the other hand, reminds you of a chamber theatre, in which a suspended DJ's console occupies the place where the director would be.

像我们所知道的一样，啤酒和鸡尾酒并不互相排斥。但这两种酒是不同的，至少它们在表现气氛的时候。因此在一个房间里将啤酒吧和鸡尾酒吧结合在一起的设计风格是一件令人却步的任务。建筑师们决定运用坚果壳原则。6米高的、中空的砖砌圆柱体将两个酒吧区分开。这些巨大的圆柱体外沿为啤酒吧提供了背景，而内部墙体环绕鸡尾酒区形成保护性的拱顶。这样就产生了两个截然不同的区，每一个都具有自己独特的氛围：开放性的、适于交际的啤酒吧。而鸡尾酒吧则为您提供隐晦的、私密的氛围。
它们的外部装修和内部设计也是截然不同的。靠近这栋建筑前部的啤酒吧占据了两层空间。一流的家具——简易的咖啡椅和暗色的橡木桌——与天花板高的钢栅栏和锈迹斑斑的柱子形成的美相互呼应。另一方面，鸡尾酒吧会让你想起一间室内剧院，在那里一个悬浮的DJ控制台占据了显著的位置。酒吧变成了一个舞台。

ANDREY GURARI

interior views
of the cocktail bar »Shot«
following pages:
beer bar »Khmel i Solod«,
general view

内部结构图
鸡尾酒吧
下一页
啤酒吧
全景图

194

ANDREY GURARI

interior views of the
beer bar »Khmel i Solod«,
details

内部结构图
啤酒吧
细部

197

Restaurant »Moskva«
Korova Bar
MOSKVA餐馆
圣彼得堡科罗瓦酒吧

Happiness Corporation
HAPPINESS CORPORATION 建筑师事务所

Restaurant »Moskva«
MOSKVA餐馆

建筑师 M. Barkhin, E. Imyaninova
位置 St. Petersburg, Petrogradskaya embankment, 18-a
面积 480 m²
完工日期 2004

类别 bar/restaurant
摄影师 A. Naroditsky

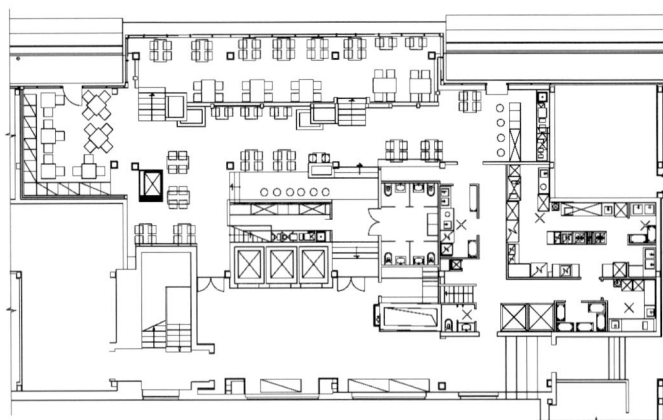

floor plan
bar with wine rack

平面图
带有酒架的酒吧

This restaurant is on the sixth floor of a business centre in St. Petersburg. The interior perfectly exemplifies the bombast of Russian neo-modernism: bright red, large fountains, heavy, baroque shapes and neo-classicising concrete pillars, which one normally associates with the Stalin period. The interior has been organised on the pattern of an amphitheatre; the main actor is outside, however, and is called St. Petersburg. For here everything centres on the fine view of the city; after all, the building is not far from the famous battle cruiser Aurora, right in the city centre. Of course you can look on the pompous decor as a concession to an elite clientele which likes to wine and dine here. Along with the bar and a restaurant area, there are a DJ cubicle and separate VIP lounges.

这家餐馆位于圣彼得堡一个商业中心的六层。室内设计完美地例证了俄罗斯新现代主义的浮华：鲜红色、大型的喷泉、笨重、巴洛克式的造型和新古典风格的混凝土柱子，任何一种设计都会让人联想到斯大林时代。内部结构依据一个圆形剧场的样子设置；然而主要的演员是在外面的，并且被称做圣彼得堡。这里每件事物都以城市美好景色为中心；毕竟，这座建筑距离城市中心的著名巡洋舰"黎明女神"号不是很远。当然您可以将华而不实的装饰风格视为对一位喜欢在此喝酒和吃饭的最上层客人的特许。连同酒吧和就餐区一起，还有一间DJ工作室和独立的VIP休闲室。

interior view:
bar with elevator and
»red zone«
details

内部结构图
带有电梯和
红色区的酒吧

203

Korova Bar
圣彼得堡科罗瓦酒吧

建筑师 M. Barkhin
位置 St. Petersburg
面积 254 m²
完工日期 2005

类别 bar/restaurant
摄影师 Y. Molodkovetz

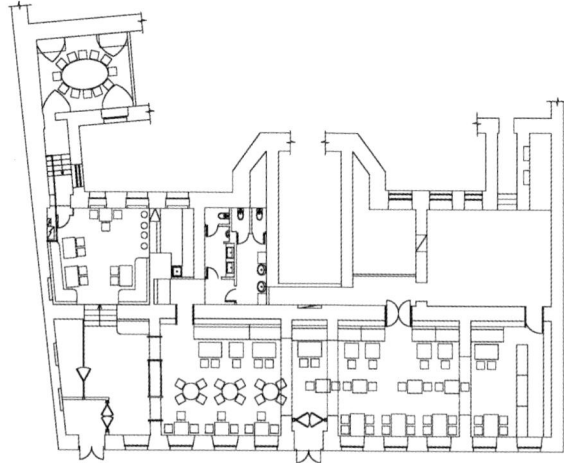

floor plan
front view of the bar

平面图
酒吧正视图

Cineastes cannot fail to connect the name of this bar with the film A Clockwork Orange by Stanley Kubrick in which there was a similarly titled milk bar. But apart from the name, the real and the fictitious bar do not have any actual similarities. Rather than milk products, this establishment emphasises the other aspect of the cow: the walls are covered with the typical patterns of the animal's hide, which in conjunction with the window sills made from polished marble, exudes an expensive, high-quality air. While animal skins appear more like trophies in other contexts, the wall coverings in this case encourage the guests to touch and feel. The room itself does not lack certain theatricality. The VIP area is positioned in the manner of a secret cabinet. Its entrance is found behind the heavy door of an antique guest wardrobe located in the lobby. This gesture reveals the idea behind the design. The clarity of the architecture unites with an almost romantic playfulness to produce a pathos-loaded atmosphere of splendidness.

电影爱好者一定会将这间酒吧的名字与斯坦利·库伯力克执导的电影《发条橙》联系起来，因为在这部电影中有一个名字类似的奶吧。除了名字以外，真正的酒吧和影片中的没有任何相似之处。酒吧内部设施着重于牛的其他方面而不是牛奶制品：墙壁覆盖着具有代表性图案的兽皮，与反着光的大理石窗框交相辉映，散发着高贵华丽的气息。兽皮通常情况下代表战利品，在这里它吸引着客人去触摸和感受。房间本身并不缺乏一定的戏剧性。VIP专区被设计成一间隐秘的密室。其入口在大厅内的古董衣橱沉重的门后。这显示了设计背后的创意。现实的建筑与虚幻的传奇轶事的奇妙结合创造出一种饱含凄美的壮丽氛围。

206

interior views
内部空间

Textiles Shop »Novator«
Restaurant »Seiji«

纺织品店 "Novator"
餐馆 "Seiji"

Iced Architects ICED 建筑师事务所

Textiles Shop »Novator«
纺织品店 "Novator"

建筑师 O. Feodorov, I. Voznesenski, A. Kononenko, M. Leikin, I. Boury
位置 Moscow, ulitsa Litvina-Sedogo, 21
面积 110 m²
完工日期 2001

类别 retail
摄影师 I. Boury, A. Kononenko, A. Yagubski

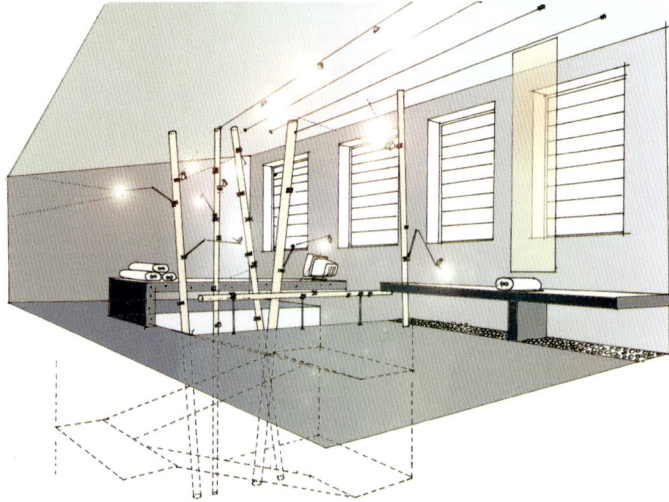

210

perspective
stair railing with
»birch tree trunks«

透视图
楼梯扶手与"桦树桩"

Everything in this shop is made from cast-in-place concrete. The entrance area, the stairs, and the sales salon in the second floor as well as all the furniture – everything is created from this hard, uncomfortable material. Idiosyncratic metal piping in the centre of the room serves as stair railings and resembles something akin to birch tree trunks in a harsh grey landscape. The shop has a total area of about 100 square metres and an unusual approach was taken in its construction. Building workers and architects worked together on-site during the building phase. The bold decision not to use detailed construction plans and calculation was made possible by continuous communication and faith in the expertise of all involved.

这家精品店的所有东西都是由混凝土浇筑而成。入口处、楼梯和二层的销售沙龙以及全部的家具，每一样物品都是由具有坚硬粗糙质感的材料制成。在房间中央的楼梯扶手是由特殊的金属管制成，它们就像以灰白色为主调的荒芜的风景画中的白桦树干一样。该精品店总面积达100平方米，建筑结构采用与众不同的设计方法：建筑期间，工人和建筑师在现场共同合作，这种不使用详细的施工图和工程计算的大胆决策通过建筑团队内部不断沟通和对技术的信心而实现。

211

Restaurant »Seiji«
餐馆 "Seiji"

建筑师 I. Voznesenski, V. Samorodova, A. Kononenko, I. Boury3
位置 Moscow, Komsomolski prospekt 5/2
面积 320 m²
完工日期 2004

类别 bar/restaurant
摄影师 E. Samorodova

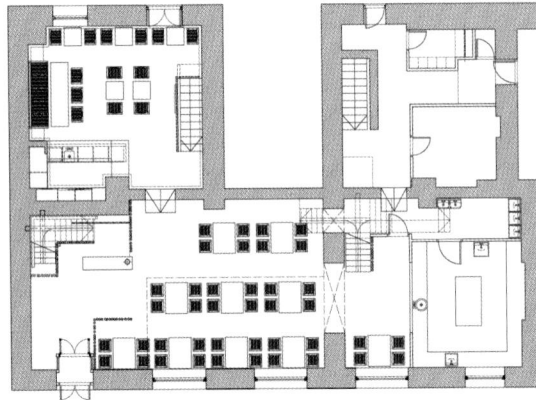

floor plan
interior view

平面图
内景

Rigid rules and their strict observance form an inherent aspect of Japanese culture. This applies equally to the interior design of restaurants. In this instance, the designers shunned any idea of copying ethnographic models. Instead they chose to be guided by Japanese literature and implemented their own conceptions of this foreign culture into the plans. The primary idea was based on the reading of the book Praise of Shadows by Junichiro Tanizaki, which deals with the ephemeral interplay of light and shadows. The natural gracefulness of the room is due to the use of materials that were taken from the Sochi area on the Black Sea: black slate, large pebbles and woods such as oak, ash or conifers. These import an atmosphere that is a combination of freshness and earthiness. Concrete pillars divide the restaurant into two sections. The wooden door separating the kitchen from the guest area is reminiscent of a typical Japanese Shoji door. Two small miniature rockeries leave the visitor in no doubt that ghosts of the Far-East reside here.

严格的规则和礼仪形成了日本文化固有的特性。餐馆的内部设计也将体现这种特性。在这种情况下,设计师避开任何一种简单复制民族学模式的想法。相反,他们在日本文化指引下,将自己对于这种异域文化的理解融入这些设计中。这种最初的想法来源于谷崎润一郎所著的《阴翳礼赞》,书中探索了阴影与光线的短暂而若即若离的相互影响。房间中大自然的优美体现应归功于从黑海索契地区运来的材料的使用,如黑色的石板、巨大的卵石以及诸如橡木、岑木和松木的木料。这些材料营造了新鲜的空气和泥土气息的氛围,会使人联想到微风轻轻拂过的小岛。混凝土的柱子将饭店分成两个区域。将厨房和客人分开的木制门让人联想到典型的日式障子门。两座小型的日式庭园无疑将会使参观者身临其境仿佛来到了祖先居住的故园。

214

small hall with wall painting
stairs, detail
interior, details

有绘画的小厅
楼梯，细部
室内景观，细部

Cherniavsky Apartment

Russian Pavilion at the 51st Biennale in Venice

Apartment in Bolshoy Kozikhinski Pereulok

Cherniavsky公寓

第51届威尼斯双年展俄罗斯国家馆

莫斯科公寓

Konstantin Larin Progress 88

KONSTANTIN LARIN PROGRESS 88 建筑师事务所

Cherniavsky Apartment
Cherniavsky公寓

建筑师 K. Larin
位置 Moscow, ulitsa Cherniavskaya
面积 151 m^2
完工日期 2006

类别 apartment
摄影师 K. Larin

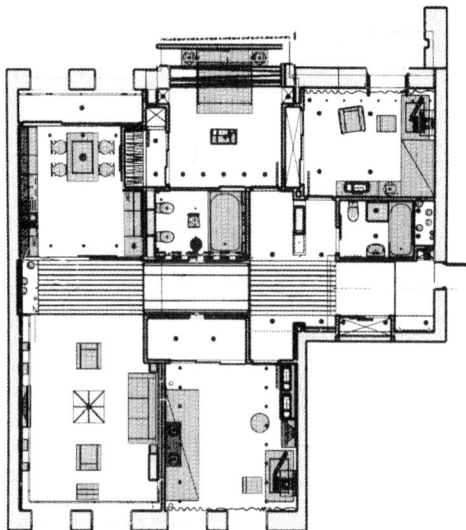

plan
living room

平面图
起居室

Architects often enjoy using new designs for living to counter stereotypes and clichés. They have a number of means at their disposal: modes of artistic alienation, or post-modern scenarios; sometimes both. In this case the architect's plan was to take a new look at Russia and its current image, to be conveyed through a combination of elements and materials which are frankly contradictory: wallpaper and oak flooring in close union with concrete, neon, steel and glass. This apartment, for a family of four in ulitsa Cherniavskaya, is at first site nothing more than a grey concrete tube, dividing the apartment into sections, each with its own level of privacy. This tube is the most characteristic feature of the spatial design. The illuminated strips running through the passage through this structure remind the visitor of an aircraft – an impression which is strengthened by the projection of a blue sky at end of the tunnel.

建筑师常常喜欢使用新的空间设计理念向老套的陈旧的设计观念挑战。对此，他们拥有多种解决方案：艺术的异化模式，或者后现代方案，有时是两者兼而有之。在这种情况下建筑师的设计图将从新的角度审视俄罗斯和它当前的形象，通过相对立的元素和材料的融合来传达：壁纸和橡木地板与混凝土、霓虹灯、钢和玻璃紧密的结合。这套为Chernyavsky街的四口之家设计的公寓，最初的选址只不过是一个灰色的混凝土管子，它将公寓分成多个部分，每一个部分都拥有独立的私密性。这根管子的特色是其空间设计。贯穿整个建筑通道上的光带会让参观者仿佛身在一艘飞行器内——管道终点处蓝色天空的影像更加深了这一印象。

Russian Pavilion at the 51st Biennale in Venice
第51届威尼斯双年展俄罗斯国家馆

建筑师 K. Larin
位置 Venice, Italy
面积 1.000 m²
完工日期 2005

类别 cultural facilities
摄影师 K. Larin

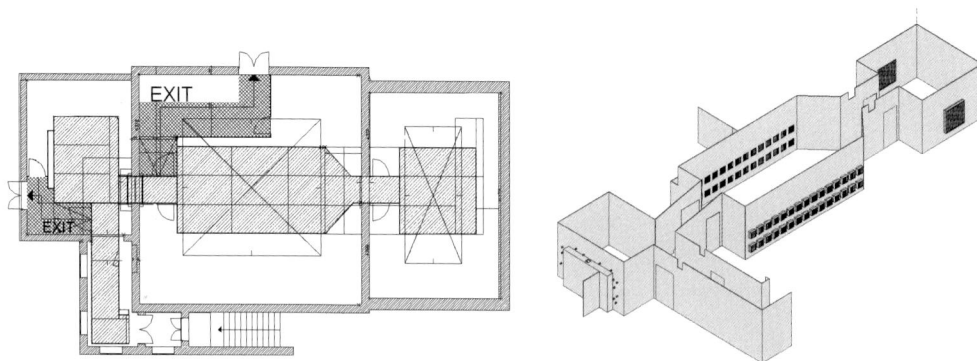

plan and axonometric view of the architectural addition to the pavilion designed by A. Shchusev interior

由A.Shchusev设计，带有阁楼的建筑平面图及景观

The air and sound installation »Idiot Wind« created by the artists Galina Myznikova and Sergey Provorov from Nizhny Novgorod was one of the exhibits on view at the Russian Pavilion during the 51st Biennale in Venice. The architecture of the exhibition is intended to coalesce with the mechanical parameters of the artwork and create a functional space eminently suitable for the specific character of this type of installation. The architect chose an enclosed dimly lit tunnel – a kind of »anti-room« within the Shchusev Pavilion. A wind machine was installed in the outer area and its ventilators fill the interior space with a howling wind. Visitors to the exhibit will, upon entering and crossing the room, find it impossible to elude the ever-increasing powerful air currents.

由来自诺夫哥罗德的艺术家Galina Myznikova 和 Sergey Provorov创作的表现气流与声音的装置 "笨风"是第51届威尼斯双年展俄罗斯国家馆的展品之一。展会的建筑设计力图与艺术品的机械参数相结合，创造了一个不寻常的有多种用途的空间以满足这类装置艺术特性的要求。建筑师选用了一个发出微光的封闭管道———一种位于舒塞夫馆内的前室。安装在外部的鼓风机将空气经通风管呼啸着鼓入室内空间。参观者进入和穿过房间时，将发现根本无法躲避开不断上升的强气流。

Apartment in Bolshoy Kozikhinski Pereulok

建筑师 K. Larin
位置 Moscow, Bolshoy Kozikhinski pereulok
面积 160 m²
完工日期 2006

类别 apartment
摄影师 A. Bogodukh

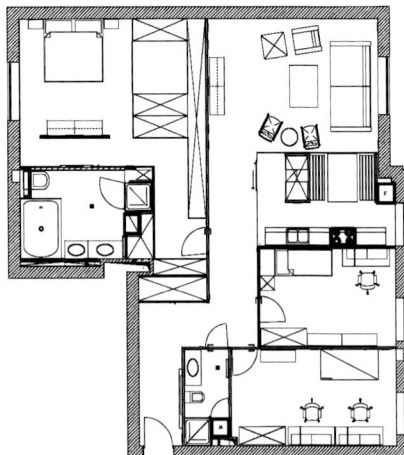

floor plan
interior detail:
pedestal light
by Ilya Piganov

平面图
内部空间
灯光基架

It is quite apparent from this apartment that the architect has experience in working with representative spaces. For it was indeed he that was responsible for the design of the Russian Pavilion at the 51st Biennale in Venice. This structure is characterised by the provision of a maximum degree of autonomy for the exhibits vis-à-vis the architecture. And it is this type of approach, whereby the radicalism is somewhat less obvious, which delineates this apartment in the Russian capital. The entire apartment resembles a wooden envelope attached to the massive building shell at only a few points – a space within a space so to speak. The warm texture of the red ash wood imputes the atmosphere with a quite cosy homely ambiance. The sober geometrical form of the modern furniture in Scandinavian style ensures the rooms are not exposed to a surfeit of rustic ponderousness, but rather lends them an airy friendly atmosphere.

从这套公寓很明显可以看出建筑师对于表象空间的设计具有丰富的经验。实际上他曾是在威尼斯举行的第51届双年展俄罗斯国家馆的设计师。这种结构给展品和建筑之间面对面的提供了最大程度的自由组合。建筑师运用这种免于张扬的激进主义表现手法设计出了位于俄罗斯首都的这间公寓。整个公寓建筑像一个木制的信封仅凭借几个接触点与巨大的建筑框架相连，可谓是一个空间套着另一个空间。红色洋白蜡木料那种让人感到温暖的色调创造出如同家一般的舒适氛围。北欧风格的现代家具所具有的柔和线条保证了房间气氛的轻松欢快而不至于过分的呆板和粗陋。

bedroom
view from the living room
to the kitchen

卧室
起居室到厨房

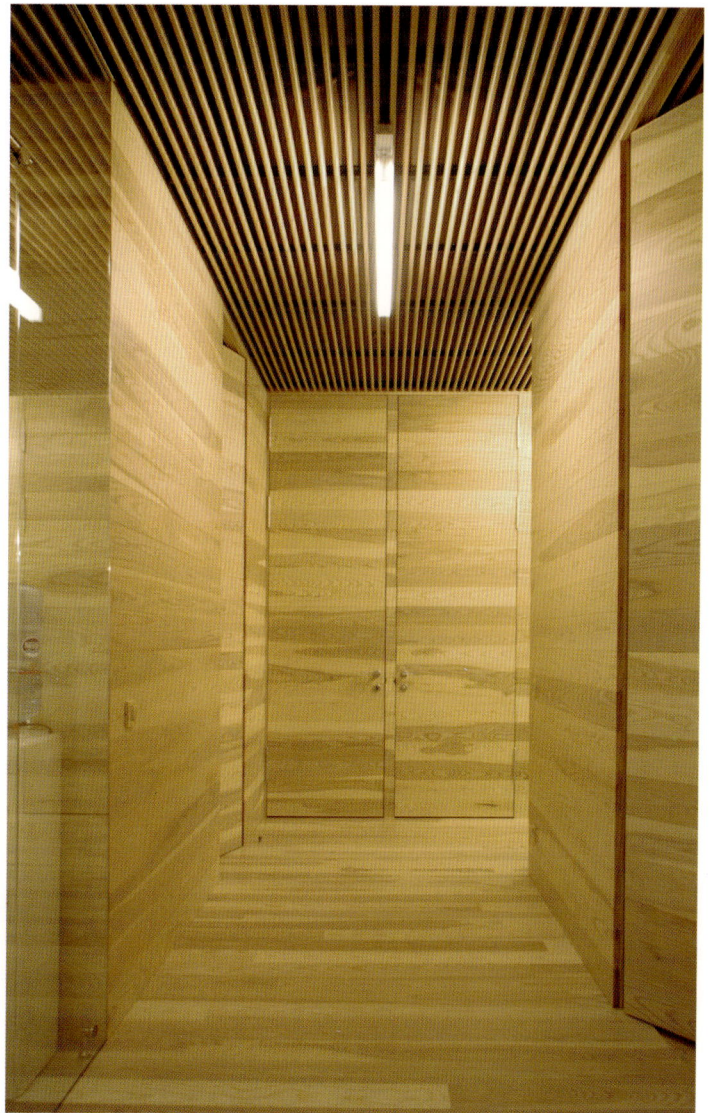

The kitchen is located
inside a glass »aquarium«.
corridor

玻璃走廊中的厨房

Apartment in Ulitsa Kirochnaya
Ozerki Perfumery
Restaurant Ki-Do
ZERO Communication and Media Agency, Cinema »Zerocinema«

圣彼得堡公寓
Ozerki香料店
日式Ki—Do餐馆
ZERO传媒代理公司

Levchuk Romanchuk Yuyukin
LEVCHUK ROMANCHUK YUYUKIN 建筑师事务所

Apartment in Ulitsa Kirochnava
圣彼得堡公寓

建筑师 A. Levchuk, A. Romanchuk
位置 St. Petersburg, ulitsa Kirochnaya
面积 110 m²
完工日期 2000

类别 apartment
摄影师 A. Strelets, A. Shinkarev

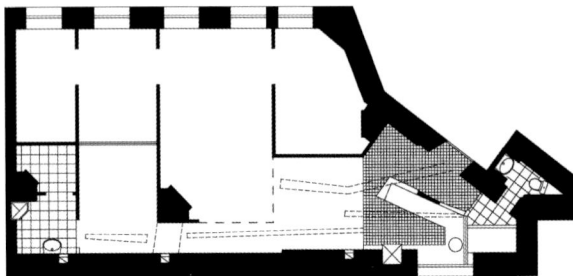

plan
living room,
antique stove

平面图
起居室
古老壁炉

The apartment is on the fourth floor of a courtyard building which dates from the late nineteenth century and has all the advantages of older, prestigious buildings. Originally the apartment formed an eight-room unit, which was used in the Soviet period as a »Kommunalka«, a communal flat. Its latest conversion was devoted particularly to restoring and repairing its detailed historical features. Thus two stoves and the ceiling stucco were returned to their original state. No change was made to the ground plan, except for separating the kitchen from the dining area and adding a new wall between the bedroom and the study. To unite the kitchen and living area under one intact stucco ceiling, the ceiling in the corridor was slightly lowered. The design won the First Prize at the 2001 Moscow Architecture and Design Festival.

这套公寓位于一处年代记载为19世纪晚期的庭院式建筑的第四层，它拥有全部较为古老的、闻名建筑的全部有利条件。最初，这栋公寓形成了一个8个房间的单位，在苏联时期被作为"公社"，一个公用的公寓 。它最近的变化是被专门贡献出来恢复和修缮它的历史特色。因此，两个炉子和灰泥天花板被恢复成原有的样子。除了从餐厅区将厨房分离，又在卧室和书房之间增加了一堵新墙外，整个公寓的平面图没有做其他的变化。为了使厨房和生活区共享一块完整的灰泥天花板，稍微降低了走廊天花板的高度。这一设计获得了2001年莫斯科建筑设计节一等奖。

Ozerki Perfumery
Ozerki香料店

设计师 A. Levchuk, A. Romanchuk
位置 St. Petersburg, Ozerki
面积 49 m²
完工日期 2002

类别 retail
摄影师 A. Levchuk

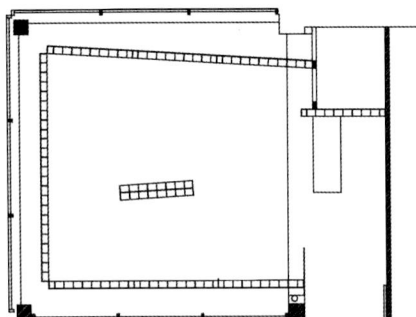

232

ground floor plan
interior

地下室平面图
内部空间

The shop is situated in a glazed extension to a hypermarket. In their design the architects attempted to utilise the transparency of this glass annex as a central determinant, exploiting it to create atmosphere. The tough shop fittings were made of 10-millimetre thick, hardened glass, the display cabinets and show cases in the shop window of 4-millimetre thin glass. Unfortunately the architect's original plan, to hang the glass walls with light and heat deflecting curtains, was rejected. Instead, a corridor for service use was made between the glass facades, now moved forward, and the stone wall at the rear. Now protection against light is provided by two large advertising banners, featuring the shop's logo.

店铺的位置设在一家高级百货商店的一处玻璃面扩建区。在设计中,建筑师尝试利用这一玻璃附属建筑的通透性作为设计的中心影响因素,用它来制造氛围。坚硬的展示架由10毫米厚的硬玻璃制成,展示橱和陈列柜放在4毫米薄的玻璃制成的橱窗内。遗憾的是建筑师最初计划悬挂玻璃墙(带有光和热驱动偏转的窗帘)被否决了。取而代之在前移的正面玻璃墙和后面石头墙之间建成了一条便于服务的走廊。现在阻挡阳光的保护设施由两个大型的带有小店标识的广告横幅承担。

Restaurant Ki-Do
日式Ki-Do餐馆

建筑师 A. Levchuk, A. Romanchuk, A. Yuyukin
位置 St. Petersburg, Kamennoostrovski prospekt, 47
面积 150 m²
完工日期 2003
类别 bar/restaurant
摄影师 A. Levchuk

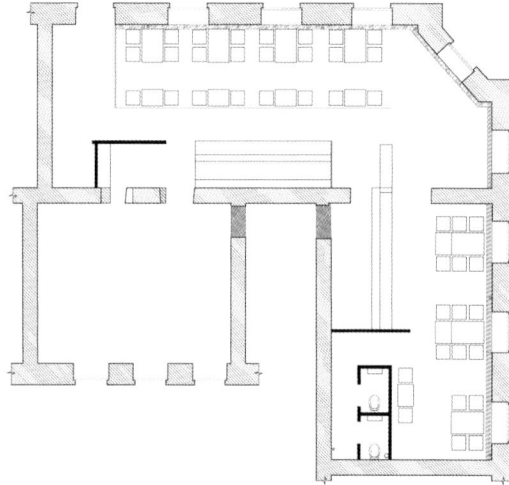

ground floor plan
interior
following pages:
general view

地下室平面图
内部空间
下一页:
全景图

It was a bold decision to resist all expectations and not to fit out this Asian gourmet restaurant in a typically Japanese style. The idea of the design is taken from the history of the Far Eastern country, where in the 16th century it had reached its apogee under the rule of the Shogun and which was characterised by the influence of the Portuguese who ruled the world's oceans at the time. With this as their background, the architects chose a rigidly occidental design. Radically enlarged Italian stylistic motifs were printed on large panels, and now decorate the ceiling of the restaurant. The spatial arrangement is based on a separation of the guests. Guests who are already dining are divided by a glass wall from those who are newly arriving. This element creates the well-known phenomenon of seeing-and-being-seen – a crucial aspect of any popular cult restaurant.

这是一个大胆的决定，它违背所有人的期望，并且没有依据典型的日式风格装修这家亚洲美食餐馆。设计的理念来自于这个远东国家的历史，那里在16世纪幕府统治时期已达到了发展的巅峰，同时也受的当时统治着全世界海域的葡萄牙人的影响。以此为背景，建筑师采用了纯粹的西方风格的设计。带有放大了的意大利风格图案的镶板现在被用来装饰餐馆的天棚。餐馆空间的安排以隔开客人为目的。通过一面玻璃墙将正在进餐的客人和刚刚进入餐馆的客人分开。这种设计创造了著名的现象，观看与被观看——这是所有时尚餐馆极其重要的一个特色。

235

236

237

ZERO Communication and Media Agency, Cinema »Zerocinema«
ZERO传媒代理公司

建筑师 A. Levchuk
位置 St. Petersburg, ulitsa Dostoevskogo, 44
面积 300 m²
完工日期 2006

类别 office, entertainment
摄影师 A. Levchuk

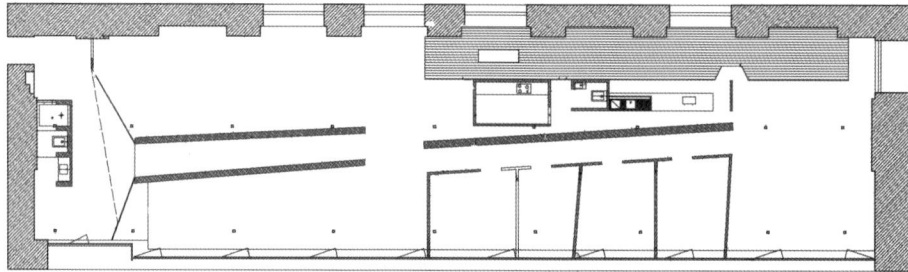

238

plan
entrance area

平面图
入口

Alongside with the standardised room schedule of a modern office area – encompassing a management section, working and office space for employees together with an open wide-area office – the owner wanted a large-sized relaxation area together with a separate media area, which could optionally be used as a cinema, conference area, photo studio and training centre. At the same time, it was required that it would be possible to accommodate events of this type alongside the normal daily business of the agency. The crucial task that was posed here therefore was to arrange the rooms in such a way that visitors and users of the media area remained unimpeded by the routine activities of the office space. And so a corridor was installed, which also serves to divide the working area into two separate zones – with the management located on one side and the creative area on the other. From the reception area, the corridor leads to the management area and the relaxation area. The use of colours is employed to denote the various functions.

一个管理区、雇员办公区和一个开放式的宽敞的办公区，除了这些现代化办公区标准化的空间安排以外，业主还希望要一个大型的休闲区和一个独立的媒体区，这里可以随意地用作电影院、会议区、摄影工作室和培训中心使用。同时需要适应代理公司每日的正常业务活动。最至关重要的任务是以何种方式安排房间，使得来宾和媒体区使用者在办公空间的日常活动畅通无阻。为此设计了一条走廊，其将工作区一分为二，管理区设在走廊的一侧，创作区在另一侧。从接待区开始，走廊通向管理区和邻近的休闲区。休闲区和办公区之间由一堵延伸的矮墙分开。不同的色彩代表区域之间的不同功能：公用设施，如厨房、洗手间和卫生区被涂成黄色；工作间被涂成鲜绿色；走廊则是墨绿色。

LEVCHUK ROMANCHUK YUYUKIN

239

various views
of the interior

内部空间多角度

240

Iceberg Night Club
White Shark Boutique
Iceberg Shopping Center

Iceberg夜总会
鲨翔精品沙龙
Iceberg购物中心

Pastushenko & Samogorov Architectural Bureau
PASTUSHENKO & SAMOGOROV 建筑师事务所

Iceberg Night Club
Iceberg夜总会

建筑师 V. Pastushenko, V. Samogorov, I. Bogoyavlenskaya
合作建筑师 A. Kononenko, P. Picshik
结构工程师 Y. Ryzhkov
位置 Samara, ulitsa Dachnaya, 2
面积 2.000 m^2
完工日期 1998

类别 leisure
摄影师 V. Pastushenko, V. Samogorov

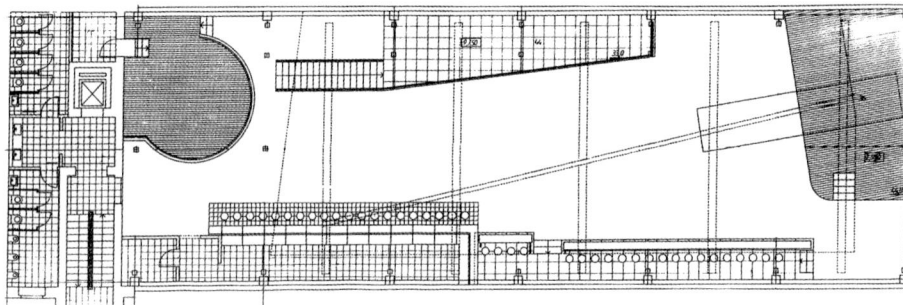

floor plan
dance floor, VIP area

平面图
舞池，VIP区

A factory is more like a machine than any other type of architecture – highly efficient, uniquely productive, with nothing superfluous, no decoration. The Iceberg Night Club uses the premises of a former factory and has preserved its industrial character. The club is an entertainment machine, and the complete invisibility of design – the obvious lack of it – emphasises the factory-like, mechanical quality of the interior. Steel pillars and beams, light metal balconies, and a wide range of technical installations resist any attempt to »do them up«, either in colour or design. Plainly the architects see themselves as heirs to the tradition of constructivism. The clear functional divisions of the interior are expressed in its vertical lines: foyer, wardrobe and offices are on the ground floor; billiard tables, bar and restaurant area, on the other hand, are one storey up. On the next storey is the dance floor, above which in turn comes the VIP Lounge. All four storeys are connected by stairs and lifts. This pleasure machine awakes at night. It produces light, sound, excitement.

一间工厂比任何其他类型的建筑更像一台机器——它的高效率、独特的可生产性，没有多余的奢侈品、没有任何装饰。这间Iceberg夜总会利用了一间工厂旧址的建筑，保留了原有的工业化特征。俱乐部就是一台娱乐的机器，看不出丝毫的设计——明显缺乏设计——重点在于类似于工厂和机械品质的风格。钢柱和横梁、轻金属装饰的阳台以及大量的技术化的装置阻止了任何将其重新整修的尝试，无论是色彩还是设计。很明显，建筑师将他们自己视为构成主义传统的继承人。按照垂直线划分的内部功能区清晰明了：休息室、藏衣室和办公室设在一层。另一方面，台球桌、酒吧和就餐区在二层。三层是舞池，在它之上是VIP休闲室，从那里您可以看见舞池中所有的表演。全部四层由步梯和电梯相连。这台娱乐的机器在夜晚醒来，为城市带来了光和声音，也带来了兴奋。

White Shark Boutique
鲨翔精品沙龙

建筑师 V. Pastushenko, V. Samogorov, K. Pikalov
结构工程师 Y. Ryzhkov
位置 Samara, ulitsa Dachnaya, 2
面积 150m²
完工日期 2004

类别 retail
摄影师 V. Pastushenko, V. Samogorov

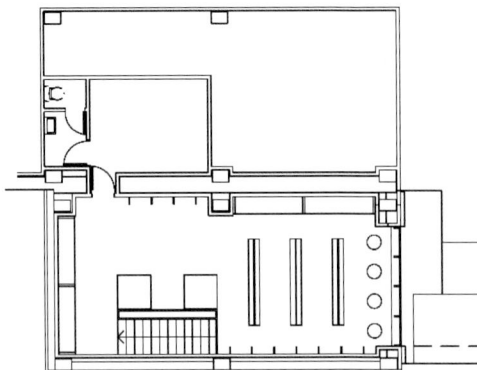

floor plan
interior, detail

平面图
内部空间，细部

Here simple clear lines are not a mere fashionable gesture, but rather are convincing and necessary. Cool rationality and self-restraint are elements these architects have already imported to numerous other projects, which they undertook as part of the development of the Iceberg Shopping Center. This boutique too forms a part of this complex. The minimalist interior more or less originates from a necessity which was allowed here to develop into a virtue. For the financial scope available for the fitting out of the shop was relatively restricted. The affordable materials – pine, simple brick and slate – were so worked that wear-and-tear from the passage of customers could be interpreted as an aesthetic statement. The quality of the architecture is not thereby diminished – on the contrary, the beauty of simplicity is part of the credo of modern architecture and is connected in no little way with the idea of honesty. Nothing here attempts to appear as something different from what it actually is.

在这里简捷明晰的线条不只是一种纯粹的时尚符号，而是合理和必需的。冷静的理性和自制力是这些建筑师已经赋予其他许多建筑的基本元素，而这些建筑已经成为Iceberg 商业中心发展的组成部分。这间小店也是这一综合建筑的一部分。最低限度的室内设计或多或少源于"必需是一种美德"的理念。用于装修店面的资金相当有限。所有采用的材料，如松木、普通的砖和石板都由于过往穿梭的顾客的使用而磨损和消耗，这也可以被阐释为一种美学立场。建筑的质量并没有因此而降低，相反，简约之美成为现代建筑理念的一部分，并且与许多真诚的创意有关。这里的一切都力求保持着一种本色的真实。

PASTUSHENKO & SAMOGOROV

248

Iceberg Shopping Center
Iceberg购物中心

建筑师 V. Pastushenko, V. Samogorov, K. Pikalov
结构工程师 Y. Ryzhkov
位置 Samara, ulitsa Dachnaya, 2
面积 10.000 m²
完工日期 2004

类别 retail
摄影师 V. Pastushenko, V. Samogorov

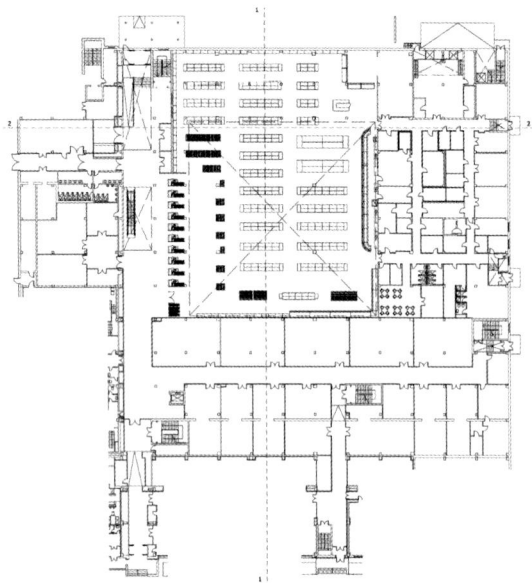

250

floor plan,
supermarket
main entrance to
the supermarket

平面图
超市
超市主要入口

The Iceberg Shopping and Entertainment Complex is a living organism, whose successful growth began with the opening of a nightclub in 1998. Since then, more and more bars and other outlets of various kinds have arrived, so that the little club has now become a veritable shopping and entertainment centre. The new shops are located in a building of their own, which features a supermarket and boutiques on its first and second floor, grouped around a light atrium. For this extension the architects utilised old buildings on the site, whose industrial charm blends elegantly with the modern design of the shops and open spaces. The daylight, falling from above, flows over the white, shining surfaces of the walls and floors; the shadows thrown by the beams trace fleeting patterns. The carelessly rendered walls and the irritating brightness of the Martini lights create an almost metaphysical atmosphere; what remains is structure, light and space.

Iceberg购物娱乐城是一个具有生命力的有机体，它的成功发展开始于1998年一间夜总会的开办。其后，越来越多的酒吧和其他各种消遣方式加入其中，因此这间小型的俱乐部现在已经发展成为一个真正的购物和娱乐中心。新建的商铺设在它们自己的楼内，环绕明厅设置的位于一二两层楼的超级市场和精品店是他们的特色。对于这一扩建，建筑师利用了原址旧有的建筑，融合了工业化魅力与时尚店铺和露天场所的设计于一体。从顶部射入的阳光在白色光亮的墙壁和地板上流淌；横梁投下的阴影追逐着飞逝的图案。粗略涂抹的墙壁和马丁尼灯刺眼的光营造了几乎超自然的氛围；留下的只有结构、光和空间。

PASTUSHENKO & SAMOGOROV

251

252

main entrance
supermarket

超市主要入口

253

interior, details
内部空间，细部

Concrete Apartment
Villa Ostozhenka
混凝土公寓
别墅

Project Meganom

PROJECT MEGANOM 建筑师事务所

Concrete Apartment
混凝土公寓

建筑师 Y. Grigorian, P. Ivanchikov
位置 Moscow, ulitsa Aviamotornaya, 10
建筑公司 Alexander Ney
面积 270 m²
完工日期 2000

类别 apartment
摄影师 A. Knyazev

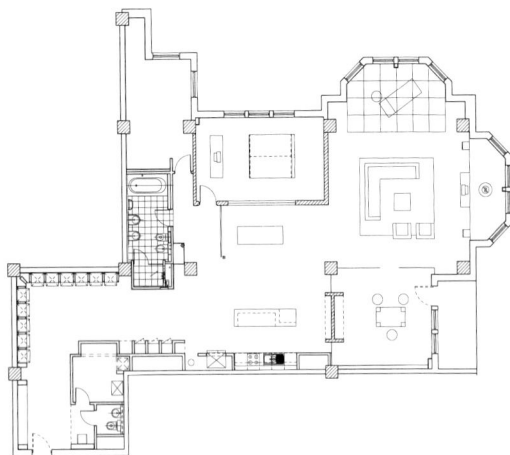

256

floor plan
kitchen area

平面图
厨房

This spacious apartment with a very complicated ground plan is situated in one of the new Moscow apartment blocks. It has two bay windows and a balcony. Given the sheer size and depth of the rooms, the best plan was obviously to organise the space around a central zone, which serves here as a kitchen and dining area. The structural elements, with their columns and beams, played a major role when it came to deciding the various room functions. Nursery, bedroom, living area and study were placed along the windowed frontage. Built-in furniture and other specially made items were incorporated into the walls. This has created a rhythmic sequence of open, closed and transparent zones, some of stone, some of wood, some of glass.

这套宽敞的公寓位于一个莫斯科新公寓楼，它拥有一个非常复杂的平面图。拥有两个外飘窗和一个阳台。最佳的平面图为房间提供了绝对的面积和高度，空间以作为厨房和饭厅的中心区环形设置。基础结构以及圆柱和横梁在决定各种房屋功能时发挥着主要的作用。保育室、卧室、起居室和书房倚着房子正面的窗户排列。嵌入式的家具和其他特意制作的物件与墙壁合成一体。石料、木材和玻璃的交替运用使得这套公寓不同区域分别具有开放的、私密的和通透等不同的特点，它们错落排列有如一曲动人的旋律。

dining room
corridor
following pages:
general view

餐厅
走廊
下一页：
全景图

Villa Ostozhenka
别墅

建筑师 Y. Grigorian, A. Pavlova, P. Ivanchikov, I. Kuleshov
建筑公司 Startex Stone
位置 Moscow, Molochny pereulok, 1a
面积 1.170 m^2
完工日期 2004

类别 apartment
摄影师 Y. Palmin

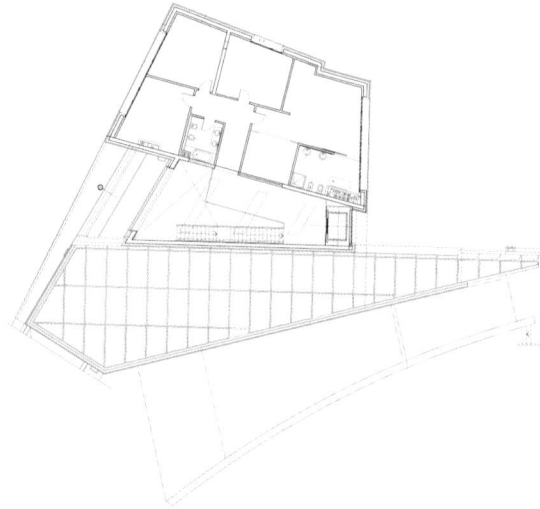

floor plan
aerial view

平面图
鸟瞰图

Ostozhenka is one of the smartest residential districts in the Russian capital. But, alas, the popularity of its location also has its negative effects. Over the last few years numerous development projects, built solely with commercial considerations in mind, have largely destroyed the district's historical context. But there are exceptions. These include this villa, whose calm reticence perfectly reflects the former spirit of the place. The building is a private, introverted refuge, whose design represents the traditional qualities of suburban houses. There is a large garden here too, and a swimming pool – though that is inside the house. The building's internal three-part structure is reminiscent of a clover leaf: the central part serves as an entrance, while on the one side lie the living rooms, and on the other the underground garage. As the visitor enters the house, he walks through a long, glass-covered hall, which ends in a lush winter garden, before he actually crosses the threshold of the villa itself.

奥斯托贞卡是俄罗斯首都最时尚的居住区之一。但是，唉，位置的优越性也产生了消极的影响。在过去的几年里，无数以商业利益为目的的开发项目，已经破坏了这一区大量的历史遗迹。但是有例外。其中就包括别墅，它毫不张扬的设计完美地诠释着这个地区原有的精神面貌。这栋建筑就是一个私密的、含蓄的避难所，它的设计代表着郊区住宅传统的品质。它也有一个大的花园和一个游泳池——尽管建在房子的内部。建筑内部的三部分结构很容易让人想起苜蓿草的叶子：中心部分是入口，而在一边是起居室，另一边是地下车库。当参观者进入房子的时候，实际上是跨进别墅门内之前，他将通过一个长长的玻璃大厅，尽头是一处郁郁葱葱的冬日花园。

PROJECT MEGANOM

263

several details
of the interior

内部空间的一些细节

内部空间的一些细节

Residence near Moscow

莫斯科附近的住宅

Inna Rannak/Elena Franchan

INNA RANNAK/ELENA FRANCHAN 建筑师事务所

Residence near Moscow
莫斯科附近的住宅

建筑师 I. Rannak, E. Franchan
结构工程师 O. Frolov
位置 Moscow
面积 150 m²
完工日期 2002

类别 apartment
摄影师 A. Rusov

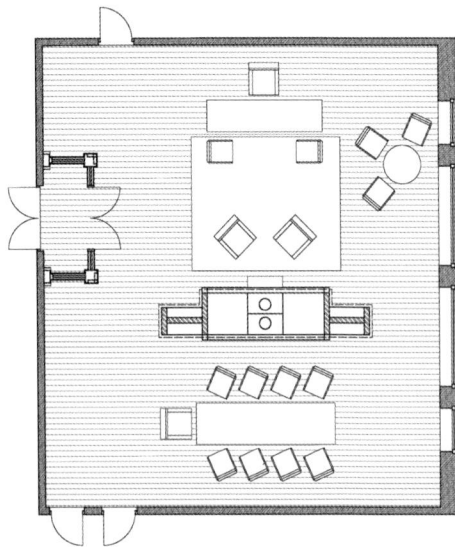

268

floor plan
fireplace and work room
following pages:
general view

平面图
壁炉和工作间
下一页：
全景

The internationally active busi-nessman has a clear conception of his Moscow residence. His work-rooms should be clearly separated from the private retreat rooms and have a different atmosphere from that of the family's domestic spheres. While the latter areas are in a decidedly feminine style, the workrooms have been staged as the masculine domain. The rustic craftsmanship of the design is related to the most original of male activities; the reference to the architecture of Frank Lloyd Wright clarifies the classical-modern demands of the head of the house-hold. Refined, warm-dark woods are used, forming a fine contrast to the representative fixtures.

活跃于国际间的商人对于自己在莫斯科的住宅设计拥有一个清晰的概念。他的工作室将要与私隐空间明显的分开，拥有一个不同于家庭内部空间的氛围。当家庭内部区域明显的体现女性风格的时候，那么工作室已经被定位于男性统治的世界。设计上的乡村技艺与最原始的男性活动相关；对于弗兰克·劳埃德·赖特（Frank Lloyd Wright）的建筑的参考阐明了一家之长的经典与现代相结合的需求。精致的、暖和的暗色木料的使用与典型的固定物形成了细微的反差。

INNA RANNAK/ELENA FRANCHAN

Freestyle Apartment
Cocoon Club
Shop »Modern«
自由式公寓
魔茧俱乐部
时尚购物店

Savinkin/Kuzmin
SAVINKIN/KUZMIN 建筑师事务所

Freestyle Apartment
自由式公寓

建筑师 V. Savinkin, V. Kuzmin
结构工程师 ZAO IPK »Bioinjector«
位置 Moscow, Daev pereulok, 22
面积 138 m²
完工日期 1998

类别 apartment
摄影师 Y. Samoshin

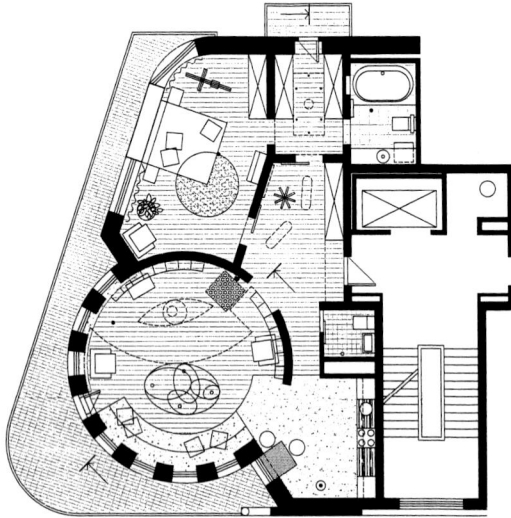

floor plan
living room

平面图
起居室

This apartment was originally a library for the people living in this large building. Following conversion of the rooms, its unusual ground plan was largely retained. At its centre is a two-storey rotunda, with a workplace in the upper storey, which you can reach from the living room via a specially constructed lift. Or you can be more energetic if you like: instead of the lift, you can use the fireman's pole at the other end. From all areas of the apartment you have access to various terraces, overhung by a belvedere directly above the rotunda. The most important stylistic element is the colour design in the rooms, by which the central rotunda and the bamboo-covered floors are welded into a uniform whole. And, as a leitmotif, a whole variety of poles will be found everywhere on the two storeys. They are both sculptural and functional elements: a hat stand in the hall, a bottle holder in the kitchen, a striptease pole in the bedroom, and the fireman's pole in the living room.

这栋公寓最初是一座图书馆，供住在这栋大厦里的住户使用。随着房间的改造，公寓罕有的平面图大部分还保留着。在它的中心区是两层楼高的圆形大厅，上层是工作间，您可以从起居室通过一个特别建造的电梯到达那里。或者您喜欢精力变得更加充沛：不用电梯，您可以使用另一端的消防队员专用的爬杆。从公寓的所有区域，您都可以到达不同的阳台，那里是从圆形大厅之上的观景楼直接悬垂而出的部分。最重要的时尚元素是房间内的色彩设计，通过色彩中心处的圆形大厅和覆盖着竹子的地板就连成了一体。并且，作为主导主题，各种柱状物在两层楼内随处可见。它们既是雕刻品也是功能元素：大厅内的一个帽架、厨房内的一个瓶架、卧室内的脱衣舞杆以及起居室消防队员使用的爬杆。

275

276

view to the rotunda
corridor

圆形大厅
走廊

work place
kitchen, detail
工作间
厨房，细部

NEW INTERIOR DESIGN IN RUSSIA

Cocoon Club
魔茧俱乐部

建筑师 V. Savinkin, V. Kuzmin
结构工程师 ZAO IPK »Bioinjector«
位置 Moscow, Gorokholski pereulok, 5
面积 360 m²
完工日期 2001

类别 leisure
摄影师 A. Naroditski, A. Yagubski, Y. Samoshin

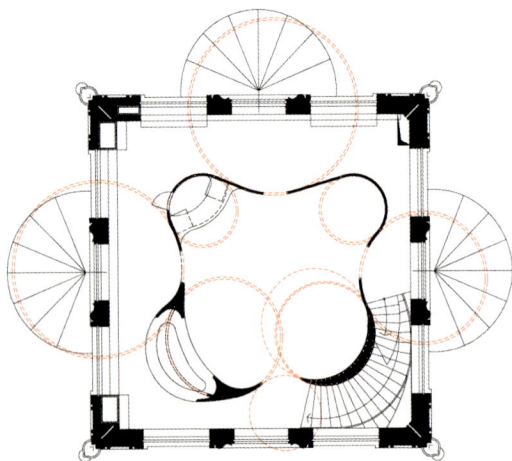

278

floor plan
interior, general view

平面图
内部空间，全景图

The complex structure of this Moscow club consists of several layers. The displacement of the external experience to the interior has a dual character: it is a semantic and temporal transition. The building itself, a relict of the recently devalued post-modern genre, does not reveal anything of its interior; it constitutes a paradox to the activity that now takes place within the rooms. Stairs made of glass and steel demarcate the passage from the past to the contemporary. The construction, which gives the club its name and is a feature running through the entire interior of the club, is a bizarrely shaped cocoon constructed from plywood. This biomorphic, technically advanced structure was one of the sensations of the last Moscow architecture exhibition, and with its sculptured object-like aura it projects an image extending beyond what one would normally ascribe to Russian design. This project sees the architecture of the nation amalgamate with international contemporary trends.

这间位于莫斯科的俱乐部结构复杂，由多层组成。外部设计经验内移具有双重的特性：语义上和时间上的转换。这栋建筑是一栋最近遭到贬值的后现代流派设计遗迹，它没有展现任何自身内在的东西：它与现在发生在内部的活动构成了一个矛盾体。由玻璃和钢制成的楼梯成为划分过去和现在的界线。建筑物本身是一个胶合板建造的异形魔茧，俱乐部因此得名，这也是贯穿俱乐部整体的一个特色。这座仿生的、具有高超工艺的建筑是上一届莫斯科建筑展览中引起轰动的设计作品之一。被雕纹装饰的仿生环境氛围，凸出的形象已经超出了任何一个熟悉俄罗斯设计人的想象。这个作品表明了该民族的建筑已经与当代国际建筑发展趋势相融合。

279

first level
lavatory, open
glass stairs to the second level,
glass stairs, detail

第一层
开放的盥洗室
通往第二层的玻璃楼梯
玻璃楼梯，细部

SAVINKIN/KUZMIN

lavatory closed
Obergeschoss
封闭的盥洗室

Shop »Modern«
时尚购物店

建筑师 V. Savinkin
结构工程师 ZAO IPK »Bioinjector«
位置 Moscow, ulitsa Tverskaya
面积 180 m²
完工日期 2004

类别 retail
摄影师 K. Ovchinnikov

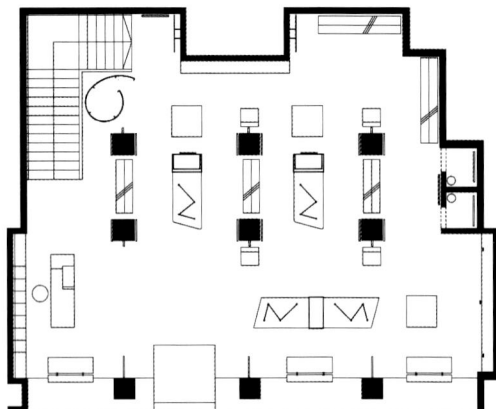

ground floor plan
shelves, detail

地下室平面图
书架，细部

This shop for young people's fashions in Moscow's Tverskaya Street is a barometer of current trends. That is due in no small measure to the interior style of this boutique, where you will find not only the finest Prêt-à-porter clothes, but also high-quality design. True, the latter is not so loud or attention-grabbing as the goods on sale. For the architect's principal aim here was subtlety – that type of subtlety through which good interior design can transform a normal shop into a quite particular space. Of course you have to look closely, sometimes even behind the facades where all the colourful clothes hang. Then you will find clothes stands made of felt, the cleverly located glass mezzanine floor, or the transparent, fragile-looking tables, kept together not by any adhesive, but by tiny metal screws. They can be dismantled and put together again as often as you like. The whole interior depends on the handmade precision with which its mathematically perfect components have been produced.

这家主要面向年轻人的店铺位于莫斯科Tverskaya街，它被视为预报现代流行趋势的晴雨表。粗略地浏览一下这家店铺的内部风格，在那里您将发现不仅有最棒的成衣，而且还有高质量的设计。真的，后者不像所售货物那样过分花哨或者吸引人眼球。在这里建筑师主要的目标是细微之处——通过那种细微的内部设计，可以将一间普通的店铺转化成为一个相当不平凡的空间。当然您必须靠近了看，有时候甚至在表面的背后，在五颜六色的衣服悬挂之处。然后，您将发现由毡布制成的衣服架子、玻璃夹层地板巧妙的设置、或者透明的看起来易碎的桌子，它们不是通过任何黏合剂而是细小的金属螺丝钉连在一起的。它们能被拆分，并且依据您喜好的样子重新组装。内部设计全部依赖于手工制作的精确性，才得以创造出这间店铺精确的完美细节。

module,
stainless steel, rose wood
show room

衣架模型
不锈钢，红木
展示间

285

Broadcasting Studio »Next and Popsa«

Apartment in Moscow

NEXT and POPSA播音室

莫斯科公寓

SL project　SL PROJECT 建筑师事务所

Broadcasting Studio »Next and Popsa«
NEXT and POPSA播音室

建筑师 A. Feodorova, A. Nikolashin
位置 Moscow, Novinski bulvar, 31
面积 850 m²
完工日期 2006

类别 office
摄影师 K. Dubovetz

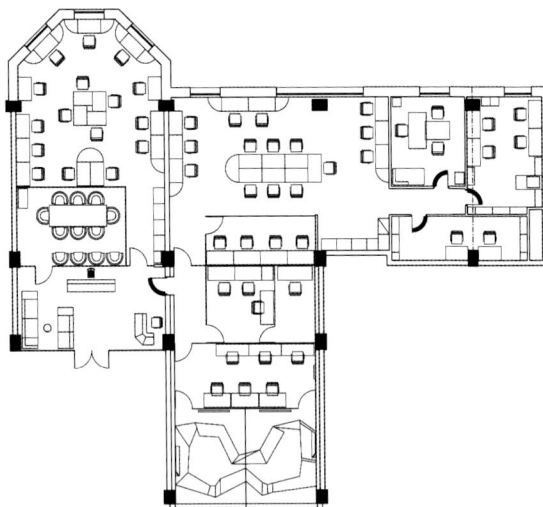

floor plan
interior
display window with the
multi-folded corpus
平面图
内部空间
展示窗

How it is possible to make the invisible visible with respect to radio transmission, i.e. the intangible connection between broadcaster and listener? This consideration was paramount in the design of these radio studios that are located in the middle of a public shopping passage. The internal area, in common with the neighbouring shops, opens up with a display window directed towards passers-by. Those walking past are afforded a view of the internal processes of a radio station: they can see how programs are produced and transmitted. An audio system broadcasts live programs into the passage. An quasi-sculptural impression is created by the whitened multi-folded corpus that, like a giant ribbon, stretches from the window front into the interior. But this element is not merely added for decorative purposes. Its purpose is at once to arrange the acoustics of the area and to connect the separate studios of the two radio broadcasters, who share the premises. Sometimes visitors sit here too, simply to listen in.

关于无线电广播（如播音员与听众之间）是如何通过无形的联接使这种无形的直观物体传播的呢？这不仅仅是无线电电波的问题：它涉及到想象力、期待，还有渴望。这种考量在设计位于公共购物长廊中部的Pate播音室时是极其重要的。其内部空间与其临近的店铺是一样的，陈列窗朝过往的路人敞开着。途经此处的人可以看到广播室具体操作的景象：他们能看到节目是如何制作和传送的。无线电中的声音被赋予了表情和真实的背景。音频系统在这个通道里播放实况录像。这些现场的声音就像书籍插画中通过变白产生的类似雕刻的效果，如同一条长长的缎带从窗前飘入到工作室内。增加这种元素不仅仅是为了起到装饰的作用。它的目的是立刻解决该区域的音响效果，并且去连接在同一建筑物内两位播音员分开的工作室。有时候参观者也坐在这里，仅仅是为了收听而已。

details

细部

NEW INTERIOR DESIGN IN RUSSIA

Apartment in Moscow
莫斯科公寓

建筑师 A. Nikolashin, A. Feodorova
位置 Moscow, Kutuzovski prospekt
面积 120 m²
完工日期 2006

类别 apartment
摄影师 E. Luchin

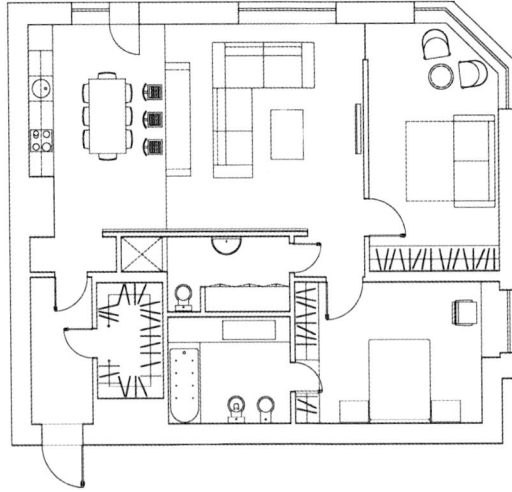

floor plan
living room
平面图
起居室

The 120 metre square apartment is located in a residential block on the Kutuzovski prospekt in Moscow. It belongs to a young woman whose budget for the renovation and extension of the apartment was limited. She wanted an open, bright room arrangement with a modern clear formal language. All rooms were organized around a wooden wall. Bright colours and glass inlays in the dividing walls create an airy, light atmosphere. The central, communicative area is bordered by the kitchen, which is separated only with a bar from the adjacent living and sitting-room areas.

这套面积120平方米的公寓位于莫斯科Kutusovsky Prospect的住宅区，属于一位年轻的女士所有。她用来翻修扩建该公寓的预算有限。她想要一个有着时尚清新的形式语言的宽敞明亮的房间设计。所有的房间环绕着一堵木质墙壁设置。隔墙上的明亮的色彩和玻璃镶嵌物营造了一个既通风又明亮的氛围。公寓中央的会客区紧靠厨房，由一个临近起居室的吧台隔开。

kitchen
The bar separates the
kitchen from the living room.
living room
bedroom
bathroom

厨房
酒吧间将厨房
从起居室中分离开
卧室
盥洗室

Jewellery Shop
Apartment in Samara
萨马拉珠宝店
萨马拉公寓

Elena and Sergey Timchenko
ELENA AND SERGEY TIMCHENKO 建筑师事务所

Jewellery Shop
萨马拉珠宝店

建筑师 S. Timchenko, E. Timchenko
位置 Samara, ulitsa Krasnoarmeyskaya
面积 400 m²
完工日期 2005

类别 retail
摄影师 S. Timchenko

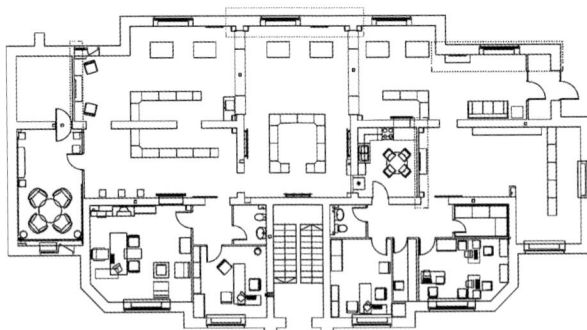

296

floor plan
interior view

平面图
内部空间

Samara is not exactly renowned for its spectacular architecture or pulsating urban life. Nevertheless, here is a shop that would be well suited as a setting for a Peter Greenaway film, and with its almost shameless opulence would be an exemplary image for any metropolis. It is however located here, on a busy urban intersection in the first storey of an inconspicuous building in the middle of a flat landscape. The arrangement of the rooms is reminiscent of a playful chateau de plaisance. Luscious coloured wallpaper, gold embellishments and decorations, Baroque furniture and heavy chandeliers leave visitors with the impression of having accidentally stumbled into a secret museum. But the observant customer will not fail to notice the details: the shining floor reminiscent of a Florentine Palazzo is composed stonework and concrete, the columns and stucco friezes are made of plaster, the seemingly authentic gold leaf is revealed to be ordinary paint. Those who cannot tear themselves away from this theatricality will stay and shop.

萨马拉，一个偏僻小城，确切地说它并不是以其壮丽的建筑和紧张的都市生活而闻名。然而这里有一家店铺应该非常适合成为彼得•格林纳威的一部电影的场景。它那露骨的富饶可以成为任何大都市仿效的对象，然而它却是位于一个小城的繁忙的十字路口处的在一片平坦的开阔地中央的一栋不显眼建筑的一层。房间的安排使人想起一处有趣的娱乐城堡（château de plaisance）。色彩绚丽的壁纸、金制的装饰和饰品、巴洛克式家具和巨大的树枝型装饰灯，使得参观者仿佛不经意间走入一间神秘的博物馆。但是细心的顾客将会发现一些小的破绽：仿佛佛罗伦萨宫殿才有的华丽地板只不过是由石头和混凝土制成，柱子和灰泥檐壁是石膏制的，看起来足以乱真的金叶儿是由普通的颜料涂成的。在这里，那些感觉自己被愚弄的人将继续前进，而那些喜欢此种夸张风格的人将继续流连购物。

ELENA AND SERGEY TIMCHENKO

Apartment in Samara
萨马拉公寓

建筑师 S. Timchenko, E. Timchenko, N. Tershukova
位置 Samara, ulitsa Sadovaya
面积 350 m²
完工日期 2004

类别 apartment
摄影师 Y. Palmin

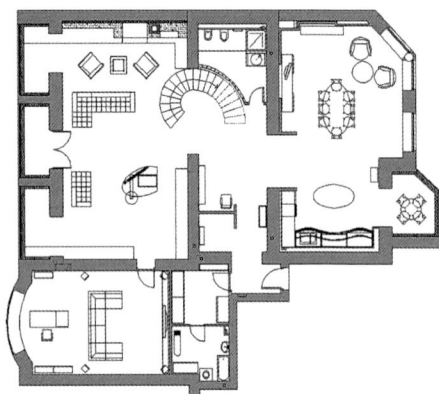

floor plan
living room

平面图
起居室

In the case of this spacious apartment, the future resident's only request was that the design not be anything remotely like classical. A German pop-jazz CD proved to be of assistance. The style, oscillating between airy electro beat and retro-club jazz, was translated into the interior design, which met the requirements of a young family with two children. Modern, light furniture combined with contemporary art established the leitmotif. The unalterable structural parameters of the simple new build posed a challenge, for its low ceilings and relatively fixed spatial arrangement could not be reconciled with the customer's ideal of loft living. To balance the deficits in this respect and to facilitate the expression of a breezy, weightless atmosphere, the rooms were painted in snow-white. Glass blocks were installed as room dividers. The 450 square metre living area is distributed over two floors: in the first floor is the living room with kitchen and dining area, while in the second are the parents' and children's bedrooms. A wide staircase flows in the vertical plane.

通常来说建筑师在设计时都不会考虑公寓主人对电子流行音乐方面的需求。但是在萨马拉这栋大型的公寓建筑设计中，未来住户唯一的要求是这栋建筑的设计不要古典得令人遥不可及。一张德国流行爵士乐CD提供了巨大帮助。那种介于空灵的电子打击乐和火箭俱乐部的爵士乐之间的风格被诠释成为室内设计的基调，这种风格迎合了拥有两个孩子的年轻家庭的需要。由著名设计师设计的时尚家具体现了与当代艺术的完美结合，奠定了设计的主旋律。由于建筑本身较低的天花板和相对固定的空间安排并不适合消费者对于阁楼居住的要求，这一简单的新式建筑因其不可更改的结构参数而受到非议。为了平衡这方面的不足，也为了尽情的表达轻松愉快的飘逸的感觉，房间被漆成了雪白色。大块的玻璃做成房间的分割墙。450平方米的居住空间分布在两层：一层是客厅、厨房和饭厅，二层则被设计成父母和孩子的卧室。宽阔的楼梯垂直设计。由于建造时设计的失误，楼梯占用了大量的空间。

ELENA AND SERGEY TIMCHENKO

302

ELENA AND SERGEY TIMCHENKO

303

Salavat Timiryasov's Apartment

Salavat Timiryasov 公寓

Salavat Timiryasov

SALAVAT TIMIRYASOV 建筑师事务所

Salavat Timiryasov's Apartment
Salavat Timiryasov 公寓

建筑师 S. Timiryasov, V. Nikiforov
位置 Moscow
面积 150 m²
完工日期 2002

类别 apartment
摄影师 M. Stepanov

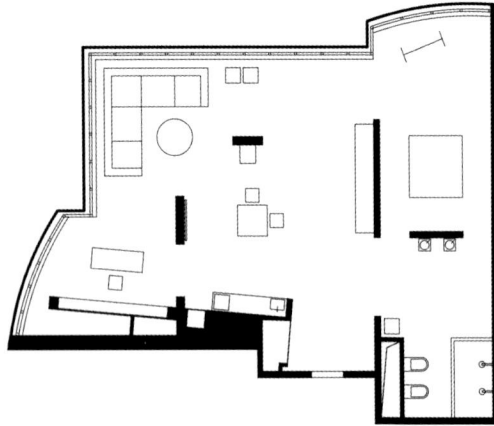

floor plan
bedroom:
sinks at the bed-head, metal
cupboards, ventilation shafts
following pages:
general view

平面图
卧室
下一页:
全景图

This generously sized apartment is the property of the owner of one of the most famous interior design firms in the country and, following a detailed article on the property in a lifestyle magazine, it is certainly now considered to be a bizarre example for the incredible settings of »new luxury« beloved of the modern Russian property classes. Precisely nine furniture items are arranged over 150 square metres – a minimalist record. The aseptic, clinically pure atmosphere is fully intentional: for the owner it was a matter of the domestication of sterility. This object is emphasised by a singular lighting concept. The background illuminated bullet-proof entrance door is the sole light source during darkness. Kitchen and bathroom allude to laboratories. Even the bedroom exhibits traits that would not be unusual in a clinic. The bed is separated from the wash-hand basin and an open toilet only by a low room divider. The illumination is controlled acoustically and – depending on the mood – can be adjusted to the futuristic atmosphere of a space ship or the cold glister of an intensive care station.

这一面积巨大的公寓的主人是国内最著名的一家室内设计公司的所有者。如果仔细阅读登载在一份时尚生活杂志上的一篇详细介绍财产的文章，你一定会发现这又是现代俄罗斯资产阶级新贵们荒唐行为的一个奇怪例证。融合怪异时尚设计风格的特殊材质的家具满足了新贵们那种故意装腔作势的要求。恰好是150平方米的房间安放了9件家具，这也是最低要求者的记录。这里故意营造出一种如同在无菌病房中一样的纯净的气氛：对于房间的所有者来说保持一个无菌状态下的生活确实是一个问题。这一项目着重以单一的照明概念为主。入口处被防弹门照亮的背景墙是黑暗中唯一的光源。厨房和浴室让人想起实验室。甚至卧室的设计特点如同普通的诊所，床和洗手盆以及开放的盥洗室之间仅由一个低矮的间隔物分开。房间的照明依靠声音自动控制，根据语调的不同可以调节出如同在宇宙飞船中未来感觉或者如同重症监护室内一样的苍白灯光。

SALAVAT TIMIRYASOV

view from the bed
to the open bathroom
living room
起居室

SALAVAT TIMIRYASOV

bedroom
various illuminations
卧室

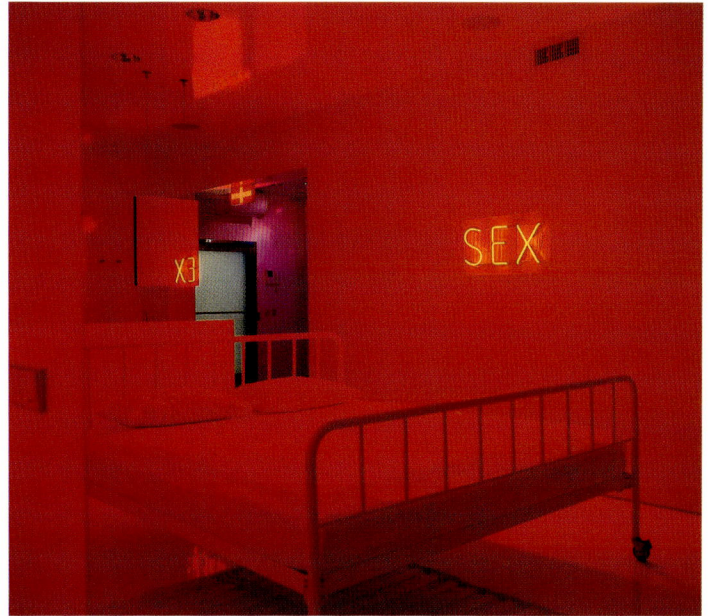

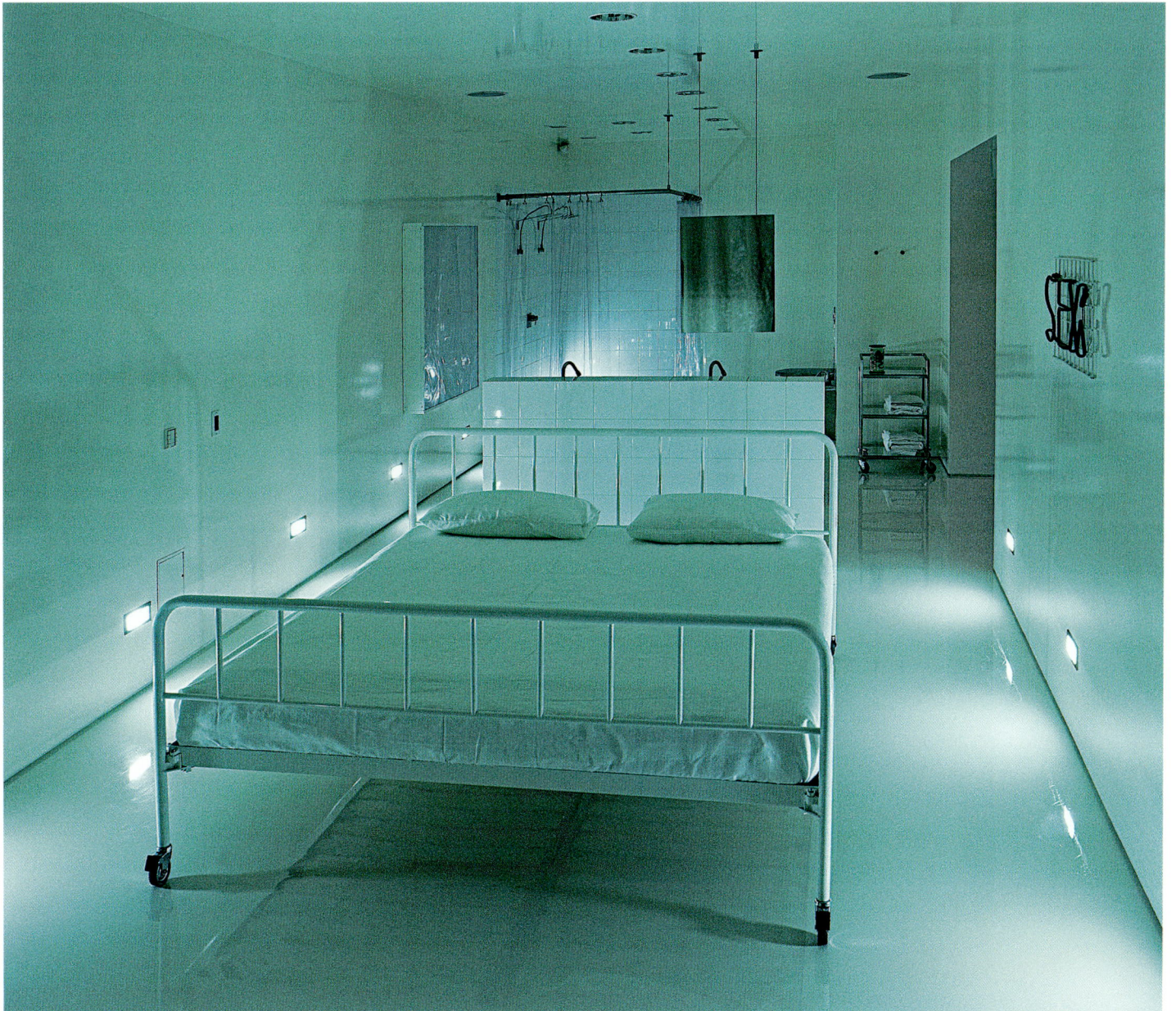

311

Apartment on Filippovski Pereulok
Conference Room of an Industrial Group
莫斯科公寓
工业公司会议室

Boris Uborevich-Borovski
BORIS UBOREVICH–BOROVSKI 建筑师事务所

Apartment on Filippovski Pereulok
莫斯科公寓

建筑师 B. Uborevich-Borovski
位置 Moscow, Filippovski pereulok
面积 110 m^2
完工日期 2003

类别 apartment
摄影师 K. Ovchinnikov

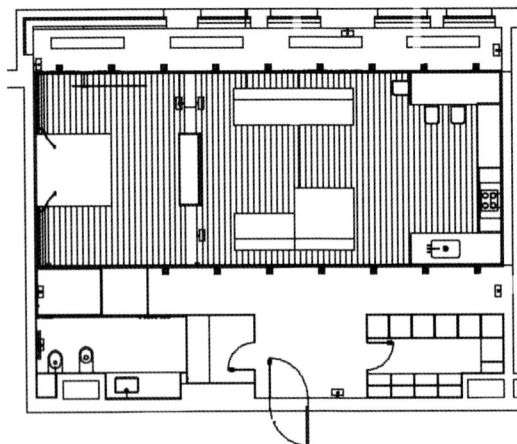

314

floor plan
living room and dining area

平面图
起居室和餐厅

The apartment is located in the fifth storey of a building in the Russian capital. The spatial area is arranged within an open floor plan and is augmented by separate rooms for dressing and closet space. The living area and kitchen are divided from the bedroom and bath by two load-bearing pillars, between which a custom-made wardrobe is positioned. A wooden shell extending in a two-dimensional arrangement along the longer side of the apartment encompasses this area, the floor surface of which is finished in parquet. Parallel to this feature there are semitransparent curtains lit from underneath. The kitchen area is located at the end of this accentuated spatial context. At this point panels made from white Corian form a U-shaped room system, which accommodates a bar alongside work surfaces, basin and hotplate. The bathroom with its black plastic coated walls is separated from the open sleeping area only by a glass shower unit and a wooden panelled bathtub.

这间公寓位于俄罗斯首都一栋现存建筑的第六层。空间分布可参见一幅公开的平面设计图，单独的更衣室和储藏室使空间增大。起居室和厨房由两个承重梁将其从卧室和浴室分开，承重梁之间摆放着定制的衣橱。沿着公寓较长的一边平面延伸的木制框架围绕着这一区域，地面铺着拼花地板。与之对应的是从下面被照亮的半透明窗帘。厨房区位于这一着重突出空间的尽头。在此处由白色的科利安制成的夹板形成了U形房间安排，沿着工作台、水池和烤盘形成了一个吧台。带有黑色塑胶墙壁的浴室与开放的卧室之间只由一个玻璃淋浴间和一个木澡盆隔开。

BORIS UBOREVICH-BOROVSKI

315

316

bedroom
space between the facade
and the wooden shell
kitchen
bathroom, details

卧室
厨房
盥洗室,细部

Conference Room of an Industrial Group
工业公司会议室

建筑师 B. Uborevich-Borovski
位置 Moscow, Ermolaevski pereulok
面积 170 m²
完工日期 2001

类别 office
摄影师 Z. Razutdinov

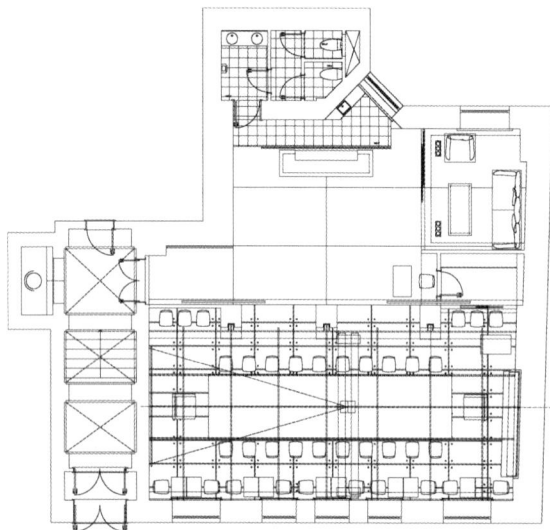

318

floor plan
general view

平面图
全景图

For the interior design of this conference area belonging to a manufacturing company the architects borrowed extensively from industrial visual and aesthetic paradigms. The entire interior here exudes the atmosphere of a large production hall; from the unsubtle iron ceiling girders to the metal floor covering to the steel table that sits up to 30 people. Piping and cabling are unveiled as autonomous spatially-defining elements; the design lends a new aesthetic attribution to materials and building components usually only found in production halls and workshops. A novel type of interior impression is thereby created in which the topos of the »Cathedrals of Capitalism« collides with a postmodern interpretation of industrial materials. The company is seeking to use this fusion of styles to best present itself within the context of new Russia.

对于隶属制造公司的会议区内部设计，建筑师广泛的借鉴了工业化视觉效果和美学范例。在这里整体的内部装修洋溢着一间大型产品厅的气氛：从粗线条的铁制顶梁到金属包覆的地板到可供30人使用的钢结构会议桌。暴露的管线形成独立空间定义元素；设计借鉴了通常只有在产品大厅和工作间才可以见到的全新的材料和建筑元素的美学属性。新颖的室内印象被创造出来。在此，繁荣时期的传统主题"资本主义大教堂"与后现代工业化材料的全新阐释遭遇冲突。这家公司正在寻求这种多样风格融合的设计来使其能够以最佳的姿态出现在俄罗斯新的发展环境中。

Vitruvius and Sons

VITRUVIUS & SONS 建筑师事务所

Architects' office »Vitruvius and Sons«
Glass Labyrinth
Shoe Boutique »Mania Grandiosa«

建筑师工作室
迷宫
疯狂之巅精品鞋店

Architects' office »Vitruvius and Sons«
建筑师工作室

建筑师 S. Padalko
位置 St. Petersburg, ulitsa Bolshaya Morskaya, 49A
面积 145 m^2
完工日期 2006

类别 office
摄影师 Y. Molodkovetz

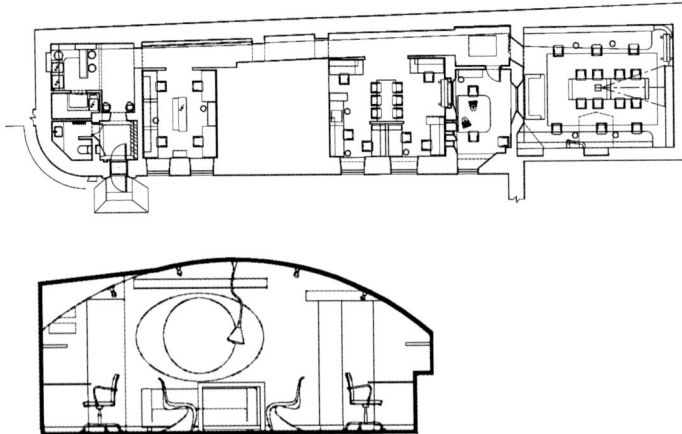

floor plan
section
workroom

平面图
剖面图
工作间

These architects' offices are situated in a spacious basement, where the walls form an attractive backdrop. A long corridor leads from the reception area to the workrooms and conference suites, and a particular feature here are the portholes in the walls. They enable you to look in and look out, thus removing the anonymity and facelessness from which office premises usually suffer. At the same time the wall between the offices is an exhibition space. Here, along with the partners' diplomas, certificates and awards, hang abstract pictures by the artist Vladimir Vasilchenko, suggested by the designs and sketches produced by the architects working here. They show the artistic roots of the buildings under creation here.

这些建筑师的工作室设在一间宽敞的地下室内，墙壁形成了一块吸引人的背景幕。一条长长的走廊从接待区通向工作室和会议室，这里最特别的是墙壁上的牛眼睛。它们使你能够向内看，也能向外看，这样消除了写字楼通常遭受到的匿名和无法辨认的痛苦。同时工作室间的墙壁构成了一个展览区。在建筑师的设计建议下，这里摆放了合作伙伴的特许证、证明书和奖品，同时还悬挂了由画家Vladimir Vasiltschenko创作的抽象画，以及在这里工作的建筑师绘制的建筑草图。在这个创作物中，他们展示着建筑的艺术之根。

VITRUVIUS AND SONS

workrooms with portholes
detail
带有舷窗的工作间
细部

corridor:
the office's library
and mediateque
detail: statue of young
Yuri Gagarin by
Vladimir Vasilchenko

走廊

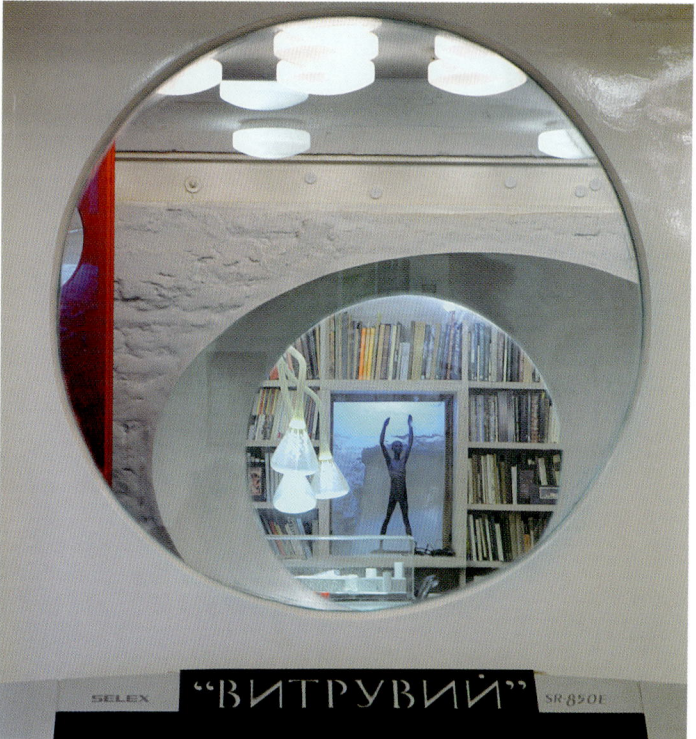

Glass Labyrinth
迷宫

建筑师 S. Padalko, A. Malkov, O. Kosenko, V. Kralin
位置 St. Petersburg, Moskovski prospekt, 10-12
面积 602m^2
完工日期 2001

类别 office
摄影师 V. Vasiliev

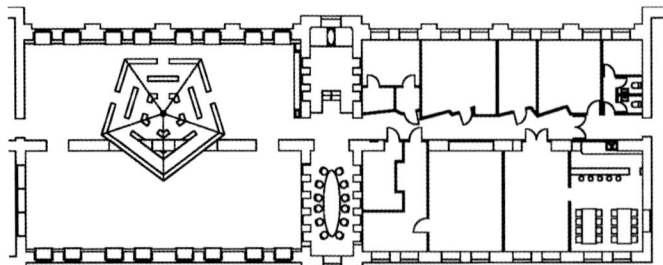

floor plan
view of the pentagonal
glass labyrinth that is used
for the exposition of the
company products

平面图

This design could serve as a perfect example of a certain type of symbolic functionalism, as developed by architects. The client insisted on a translation into structural form of the literary mythology developed in the Amber Chronicles of Roger Zelazny and not only wanted to construct the Temple of the White Unicorn, but wanted a meeting room too. After a good deal of reading, the architect suggested grouping the rooms around a central exhibition area – in the form of a glass labyrinth. This labyrinth is a pentagonal ceiling of matt glass, beneath which stand glass display cabinets featuring amber products. Given the manifold interpretations possible of this highly symbolic five-cornered shape, it seemed obvious to use the same geometrical figure in other places too – wherever that was possible. Staff join together in the Temple of the White Unicorn for festive occasions.

这一设计可以被视为设计师创造的某种符号功能主义设计的完美典范。客户执意要求将罗杰•泽拉兹尼著的《安伯编年史》表现出的文学奇幻风格诠释在结构的布局中，他不仅想要建造一座白色的独角兽宫殿，而且还要一间会议室。经过大量的查阅文献，建筑师建议围绕中心展览区布置房间——以玻璃迷宫的形式。五角形的天花板由麻面玻璃制成，它下面玻璃展示柜摆放着特色的琥珀制品。这种具有高度象征性的五角形被赋予了种类繁多的设计创意，很明显在其他的只要是有可能的地方也使用了这种几何图形。员工们聚集在白色的独角兽宫殿里举行盛大的庆祝活动。

general view
of the exhibition hall
detail

展示厅全景图
细部

»Temple of the
White Unicorn«
corridor
走廊

Shoe Boutique »Mania Grandiosa«
疯狂之巅精品鞋店

建筑师 S. Padalko, A. Malkov, D. Khimsheeva
位置 St. Petersburg, Nevski prospekt 41, ulitsa Italianskaya 15
面积 90m²
完工日期 2006

类别 retail
摄影师 Y. Molodkovetz

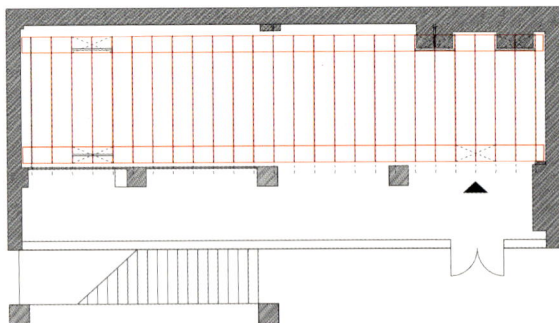

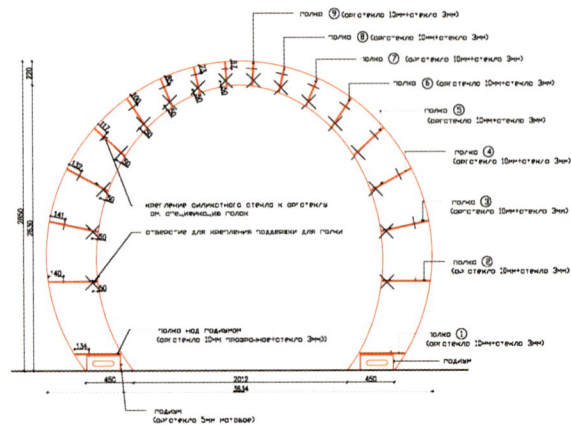

332

floor plan
section
interior view

平面图
剖面图
内部空间

The Nevski prospekt in St. Petersburg is one of the finest addresses in the country. Situated along the avenue is the Grand Palais, which houses many designer stores including this shoe boutique. In contrast to its neighbours with their somewhat aristocratically pale white and pastel-colours, the store excels in its dramatic black and red design and expressive spatial arrangement. The shop itself is designed in the form of an elongated tunnel, the curvature of the ceiling rising over long shelves that extend into the depths of the interior. The two ends of this long passageway are completed with room-high mirror installations, which interact with the glazed display areas on the long side and the shimmering ceiling surfaces to produce a provocative optical prism. The start and the end of this tunnel are thereby made wholly indiscernible. Visitors who dare to enter this void lose contact with the real world surrendering themselves to a glistening confusion of reflections, fragmented images and deceptive mirroring effects.

圣彼得堡的 Nevsky Prospect 是这个国家最有特色的主街道。大皇宫位于这条街上，街道的两侧分布着众多设计师店，也包括这家精品鞋店。与邻店相比略显有些贵族的苍白气质和柔和的色彩，店铺的优势在于它生动的黑红设计以及富有表现力的空间设计细节上。店铺内部被设计成为一个延展的隧道形式，曲曲弯弯的天花板高悬于长长的货架之上，直至隧道尽头。长长的通道两端安装了齐顶高的镜子，与两边光滑的陈列区和微光的天花板表面相结合创造出了一个具有趣味性的光学棱柱现象。隧道的起点和终点难以分别。敢于进入这一空间的参观者将顿觉与现实世界失去了联系，臣服于反射、破碎的图像和欺骗性的镜面反射效果形成的迷幻世界。房间的实际尺寸与它难以分辨的起点和终点一样不可测量。

333

334

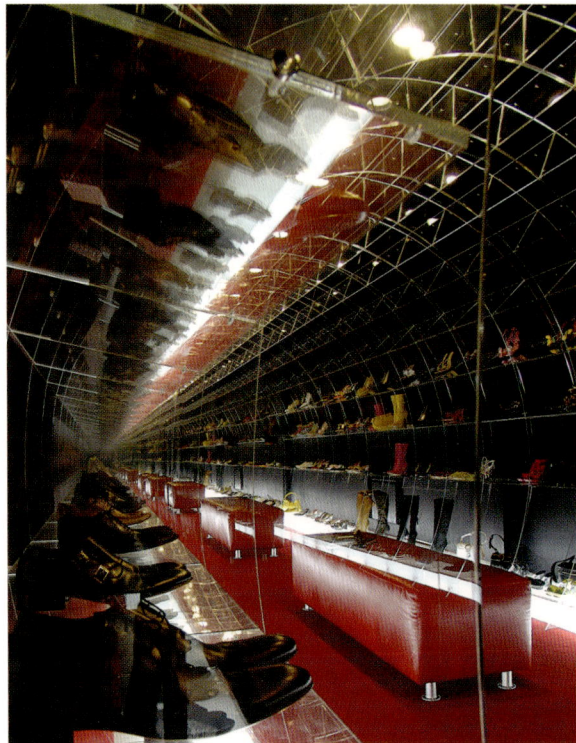

interior views
details
内部空间
细部